T0208326

Lack *and the* Art *of* Deprogramming

A Brief Look at the Universe and Our Mental State from Inside the Human Mind

Marcus William Morris

authorHOUSE®

AuthorHouse™
1663 Liberty Drive
Bloomington, IN 47403
www.authorhouse.com
Phone: 833-262-8899

Published by AuthorHouse 04/24/2023

ISBN: 979-8-8230-0587-6 (sc)
ISBN: 979-8-8230-0588-3 (hc)
ISBN: 979-8-8230-0589-0 (e)

Library of Congress Control Number: 2023906639

Print information available on the last page.

This book is printed on acid-free paper.

CONTENTS

INTRODUCTION

In the moments, the idea to write this book came to mind; I experienced a lack of love and a pang of deep sadness in my life. This profound sadness in the center of my being was becoming increasingly difficult to shake. I was receiving these feelings from within that I wanted to go home. But this was not a physical home, but a home where I am emotionally free of the pain and symptoms of Lack. This home is full of love and infinite abundance.

At the moment I decided to write this book, I was struggling to find abundance and love within myself. I know that it is there because I have experienced it before. It is that from which I came, the origins of us all. Sometimes I can achieve some level of happiness or non-sadness. Then there is just that something in my day that turns my mood downhill all over again. Be it something that I read in the news, financial and personal relationship woes, etc.

One busy afternoon, upon entering the dank and dirty subways of N.Y.C., I step onto the D train and suddenly had a jolt go through me, as my mind was bestowed with a lifetime of information all in one shot. I had this epiphany that everything in this Universe down to the smallest nanoparticle is LACKING something. So, in turn, every last nano molecule in the Universe is attracted to something else to fill in its void.

I could imagine the creator of all existence in the very beginning must have said ("let there be lack" so that every one of my creations may need another one of my creations to sustain itself). That way, the Universe had its order and remained whole. Instead of splitting into ever smaller disconnected fragments of what once was whole. Things are meant to come together in a puzzle-like fashion. A great example of this is our D.N.A.; our D.N.A. is made up of proteins (A, C, T, and G) coded for specific functions to work in unison in creating a living breathing Human Being. A protein by itself is useless.

Our Human species seem to have been exacerbating our experience of Lack. So in retrospect, with all the vast intelligence we are endowed with, we still wage wars and destroy our environment on a boggling scale. Thus this Lack, which I have observed to be one of the universal laws governing our shared existence, is focused on Humankind with the intensity of a child holding a magnifying glass over ants, and on some people more so than others.

While thinking to myself in deep concentration or meditation, it came to me that it isn't anything but thoughts that cause my suffering. It is my internal dialogue, which is giving me my perception of the world. Don't I have control over my mind? If not me, then who else? I couldn't possibly think of anyone or anything that would have control over the private thoughts racing through my skull but me. Every regressive idea that has ever come into my mind could only exist because I have somehow allowed it to be there, and I confirmed their existence by repeating them over and over again in my mind. As a result, a neurological four-lane highway is created in that glorified fatty tissue of mine called the brain. So, in turn, I experience these crazy counterproductive thoughts that torment my very existence. Sounds slightly sadistic, huh? But we all do it. Ask anyone!

Plain and simple, we have to take control of these processes going on in our minds, we are the ones thinking, and no one else. All we have to do is increase our self-awareness and pay attention to these regressive vermin (self-inhibiting thoughts). If each human being did his or her part to keep their reactionary thoughts in check, our collective intelligence works together in a symbiotic relationship. This would be enough to turn this world into something that looks like a heavenly Utopia compared to what we see today. In today's age of technology, and with the knowledge of our current resources, there has never been a better time to get started. So why have we not done this already? It's because we all have personal issues of magnified Lack to overcome in our lives, creating varying levels of disorder and malfunction.

From the infants' very first cry to today's current wars, these blatantly obvious symptoms of Lack can be dissected and traced to their leading underlying cause. In discovering what the fundamental Lack may be, you will likely encounter the solution. I believe this to be a simple truth. But instead, man loves to find a way to complicate everything. Even the very air we breathe and the water we drink is complicated through the use of chemicals and pollutants, and the same goes for our food.

According to my observations, this phenomenon seems to be self-evident. With their five senses, any person can observe these events throughout history and time to conclude. I just merely state what has always been there.

So what is the purpose of this apparent magnified Lack that we all feel? Why does our Universe seem to be designed this way? And how can we rediscover the infinite abundance within and reclaim the experience of being a satiated, whole, and loving being?

CHAPTER 1

THE BEGINNING!

Some people believe that the Universe was created by some kind of extremely fantastic explosion of one isolated singularity in the middle of a formless, empty void. The event is something quite unimaginable, and anybody with a brain can put two and two together, and see that this couldn't possibly be the beginning, because I'm sure many of you know that something cannot come from anything. The exploding singularity, and the "empty" void that it was in, had to get there somehow. Quantum science clearly states that nothing is genuinely empty. This means there would have had to be some sort of existence before this great event. Exploding singularities with the energy and mass to create an entire universe cannot just manifest into space without a catalyst.

Other people believe in a more biblical version of creation, where God uses his words and his voice to create the heavens and the Earth in the Book of Genesis. The three major religions, such as Judaism, Christianity, and Islam, share this common belief. Christianity and Islam both stem from Judaism, and Judaism directly comes from Zoroastrianism, the first monotheistic religion. However, their accounts of human creation and cosmology have their differences. According to the Zoroaster religion, the first human beings were hermaphrodites. But since most people are unfamiliar with the Zoroastrian faith, I will not discuss it further. I will leave you, the reader, to do your research on that subject.

In Genesis, God Creates the Heavens and the Earth in Seven days through his use of words and the power of suggestion he instills upon the universal matter. He speaks into a formless void and out springs the Earth and Heavens. The Earth was a water world steeped in darkness. Then God spoke the words, let there be light, and there was light. Then God had to do something about all the water present. So he created a vast space in between the waters. There exists a water canopy in the Earth's atmosphere. Beneath the water canopy, there was the sky. This is what was meant by above and below the firmament. Water was gathered together to create the oceans that we have today. It did not rain in those days. Water came up from springs in the ground. Here is one interesting detail; God made day and night when he had said let there be light on the first day before he created the sun and the moon, which was not created until day four. How does this work out? The Bible gives no answers to this.

From then on, God creates everything else, and "they," God says (us) explicitly without specifying who were the others present during this creation event, creates man in His likeness and image, as male and female. The first male named Adam is created from the dust or clay of the Earth.

I'm sure all of you reading are familiar with the story of how Eve is created out of Adam's rib. This would make Eve, Adam's "female" clone. Adam and Eve were placed in eastern Eden, where God had planted a garden. There were two significant trees in the Garden; one was the tree of knowledge of good and evil, and the other was the tree of life. A river was in "The Garden of Eden," which was split into four smaller rivers. Gold and onyx were found by the first river called "Pishon," and God said the gold and onyx were good. The Bible also clearly states that man is put there to work the ground and take care of it. It seems that man is made to be a miner of gold and onyx since God had placed man in a region to work the land where these minerals are present.

Before Adam and Eve ate from the "Tree of Knowledge of Good and Evil," they seemed to have been in a perpetual state of stupor or low awareness and had no idea they were naked. I think there is a difference between someone who knows that they are naked and does not care, and someone naked, and has no awareness that they are, (similar to the state of mind of an animal) From what I observe through the text. Humans in this state would have made the perfect slave due to their Lack of self-awareness and Lack of knowledge of good and evil. Due to this lack of knowledge, Adam and Eve had no way to develop a moral understanding of deeds and behavior. So anyone with superior knowledge would have easily been able to manipulate Adam and Eve without ever having to delve into reasoning.

The Serpent slithered into Eve's presence and convinced her to eat from "The Tree of Knowledge of Good and Evil." The Serpent told her that she surely would not die, but instead would be able to tell the difference between good and evil and know things she has never known before, and become like God. So she ate it, liked it, and gave some to Adam. This angered God. The Lord God said, *"Man has become like one of us."* (Once again, "us" is used, and the Bible gives no mention as to who the others are) God continues with, *"Man can now tell the difference between good and evil. He must not be allowed to reach out his hand and pick from the tree of life and eat it. If he does, he will live forever." (Genesis 3: 21)* After that, God drove Man out of the Garden to work the land he was made of.

After reading this, I realized, according to the book of Genesis, it appears that God did not want Man to know the difference between good and evil, and to become enlightened overall like God is. Nor did he want Man to live forever after attaining this state. So man was condemned to return to the dirt from which he came.

Now that I have enlightened myself through the shock and awe I experienced while reading "God's" word. I realize that either the Bible is purposely written to place a veil over my eyes so that I may not know the truth. Or the Bible is written for someone with the intelligence to realize that the Biblical Deity described as God could not possibly be the creator of all things in existence. Could the God of Biblical text have a hand in the creation of man as his perfect slave? I think it's possible, but I'm not sure. Maybe this is why God seems to have such an incessant need for man. Some view the Biblical God as a Demiurge.

But here's a third suggestion, maybe only those who are initiated into closed secret orders have the rest of the knowledge available to them to understand what is going on in the Bible. No matter what the truth of the matter is, it's challenging to prove anything, either way, dealing with an ancient text. The answers we have conjured up dealing with human origins and the creation of the Universe are mere speculation, and ancient writings are full of discrepancies.

The creation of the heavens and Earth, along with the ever-so-consequential fall of man is indeed a lovely story. Another story, just as enamoring and deep, is the fall of Lucifer. God's most beautiful and beloved angel, how tragic! You see, God loved Lucifer and used Lucifer to cover him with his beauty. God decorated Lucifer with the most precious of stones. He was called "The Morning Star," the bringer of light. This made Lucifer feel proud and unique. So here is where the conflict comes in. God creates man out of the clay of the Earth. This man of earth elements was the pinnacle of God's creation. God was so proud of this creature that was fashioned after himself, that he made all of his angels, including Lucifer, bow down before Adam. But Lucifer was angered by this and refused. Lucifer thought why should he bow down to man when he was created first? He felt that man should bow down and worship him instead. This pride and conceit angered God, and God cast Lucifer out of heaven along with the angels who followed him.

After Lucifer was cast out of heaven, his first goal was to corrupt the heart of man. He was jealous of Man receiving God's glory before him. So he came down into the Garden and convinced Eve to eat the Forbidden Fruit. Lucifer's goal has been the total and utter destruction of Man ever since.

The eating of the forbidden fruit may have also been the development of the Ego in man. The light and dark, the yin and the yang. But what does this all have to do with Lack? We have a lack of truth and understanding of who we are, where we come from, and where we are going. Depending on what culture you are born into, and what information you have been exposed to, people are bound to have vastly different views on these subjects.

So now let's take an even closer look at the beginning of Genesis, and possibly draw some substance out of it with the clause of "Cymatics"; the study of vibration and wave phenomenon. According to Genesis, we are left to conclude given no further explanation about the creation process besides God speaking things into existence, that this process must be Cymatics. Hence sound = vibration and wave phenomenon. Every object has a frequency, meaning a rate at which it vibrates; this includes living and non-living things. If the frequency changes, the form of the life or non-living material also changes.

So let's say you have a thin piece of cellophane, and you place some talcum powder on top. Then you create a frequency through a sound source with the use of vowels, let's just say you start with A, then E, I, O, and U, as you change the vowels, the geometric shape on the cellophane will change. The higher the frequency, the more complex the configuration will be. The level you can go to with this experiment is nearly endless; you can even create things that look like planets and galaxies, and add interference frequencies so that the substance develops a rhythm and pulsates like a heartbeat. But once the frequency ceases to act on the object, its form is destroyed. In the beginning, it is said that the Earth had no shape or form. So God spoke and gave it form. Whether or not the science of Cymatics was the Bible's intended message is not of my knowledge. But there it is, and this science was also presented by C.S. Lewis, the author of "The Chronicles of Narnia," in his first book titled "The Magicians Nephew." A character in the book named Digory puts on a ring, and upon doing so, slips into a parallel dimension where he finds a Lion using his beautiful voice to create a world and all of the creatures in it, and out they spring from the ground in full form. Another interesting parallel is that this world was created to be free of evil, but somehow evil found was let in...

So is Cymatics a new science or an old science? Well, there were undoubtedly others in our past that seem to describe its workings or give some hints to its existence, while not fully disclosing how it relates to everything around us and our very own life. Some ancient cultures claim to have used sound to lift megalithic stones into the air weighing countless tons. Though I have no current evidence to prove how this may have been done, why would ancient cultures keep records of this happening without there being any basis in truth? Certainly not to entertain their children with fantastic stories, or to deceive those living today. The fact that I am talking about Cymatics right now proves that all knowledge of it is not lost, but instead severely suppressed.

I am sure many of you are thinking, why would scientific minds suppress such an excellent and fundamental working of our very existence in the Universe? Knowledge is power, and if everyone has the information, it will empower people, and it is tough to dictate to empowered people. Upon observing experiments done with the use of sound or vibration, you would see intricate geometric shapes develop as if a higher intelligence created them. All sorts of shapes and forms develop that seem to reflect the Universe and the world around us. You will realize there is no such thing as real chaos and randomness, and that everything obeys a universal order, an uncanny "intelligence". You will discover sacred geometry, and you will see this reflected in absolutely everything. The Universe is truly holographic. For instance, take a look at a model of an atom. Then check out a model of the solar system; the same form is present. The difference is in scale. Extraordinary huh? The clues are all around us. All we must do is open our eyes and see. Awaken to the truth before our eyes; it is not hidden.

Take a look at a hurricane then look at a galaxy. Look at the blood vessels in your arm then a tree. Have you ever heard of Laminin? It is a protein that holds every one of our cells together, and here is how it looks.

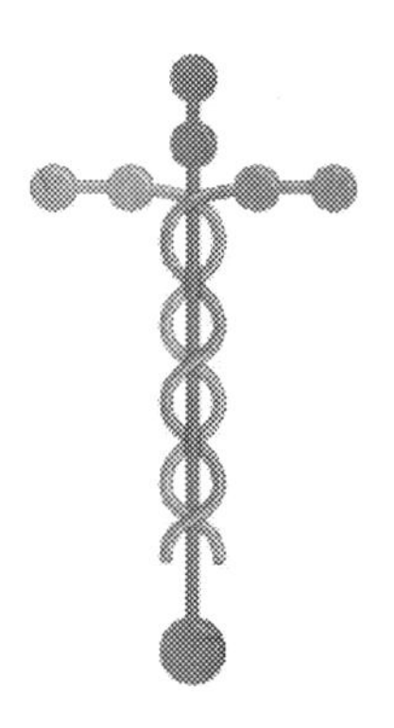

Could it be just a coincidence that it is a perfect scaled-down holographic model of the human body and also follows the measurements of sacred geometry? 1, 1, 2, 3, 5, 8, 13, etc, or 1.680339887 the golden ratio. The Golden ratio was maintained as heavily guarded knowledge, which was in use during the building of the pyramids and passed in secret to include the building of the world's cathedrals. The structures left behind built with the use of this information are extremely advanced compared to the adjacent Stone Age structures built by the average man. I guess it also may be a coincidence that this looks like a cross or the crux, a very ancient pre-Christian symbol. The Egyptian example of this is called the Ankh. So I would have to say that in the beginning there was sacred geometry and there was also sound.

The beauty of sound, nothing I ever heard sounds like it. The creator must have sung the Universe into being along with all things in it like the lion of "Narnia," Aslan. The more I learn about this world that I live in, the more beautiful and mysterious it becomes. I find myself unlearning many things that have been ingrained in my mind since birth. This long life journey leads me closer and closer to some things and further away from others.

The quest I am on is a mix of great pain and pleasure. I endure piercing loneliness, knowing that very few people understand my interest, and even fewer people to talk to about them. On the other hand, I am wallowing like a pig in a pen of forever renewing knowledge. What brings me comfort is that I am being myself and relearning parts of me that I have long forgotten. When I change for others and become what they want me to be, I lose myself and can never indeed be happy. I can only achieve happiness when I am me. Now that I know the thoughts that I am thinking, are the thoughts of my choosing, I will try my best to learn to dissipate what I think controls me. So in this quest, I will learn to achieve that. We will learn to accomplish that.

CHAPTER 2

BIBLICAL GOD & LUCIFER

What's evident in all the world's major religions is that God seems to desire worship and the obedience of Humans, and if people fail to cooperate, God seems to hand out dire and deadly consequences. Why is this the case? Is there some sort of incompleteness to the "God" the world's religions worship? Why would an all-powerful, all-knowing, omnipotent being desire the attention of massively inferior Human beings with their limited knowledge and power, which is infinitely smaller than the creators of this Universe, and possibly multiple universes? Could it be possible the creator of all things in existence may have some sort of need for Humans as if people can offer the creator some kind of much-needed help and attention? In my personal opinion, I don't think so.

The "God" of the Bible and all other religions seem to express a very incessant lack, an all-consuming need to control the will and desires of Humankind, a need to steer humans on a path most desirable for solely himself. Humans in the

Bible and other holy books are treated more like a commodity to be controlled rather than precious ever-loving emotional beings, with wants and needs. For instance, there are many examples of God destroying human beings when they do not act under his will. Would you kill your children, your best friend, or your spouse if they disobeyed you? Most people would not. God is described as a perfect being and all-loving in every way, and at the same time, regularly kills his creation and demands sacrifice. In my eyes, this makes absolutely no sense. A God of love could easily shine his passion into the heart of any situation to heal. But the God of the Bible, in many cases murders instead.

I just don't get it! Most of us who are familiar with Biblical texts know that we are given free will. But I do not see why a God who could come down and heal through love, use fear and murder as tools for control instead. I'm sure that direct healing through love would have the effect he wants, without affecting free will any more than murder would.

Personally, in my eyes, you could either be a God that is forever unconditionally loving, choosing to love humans through all flaws he created them with, or a God who rules through murder and fear and promises hell if you disobey his command. I don't think you could be both! But my argument is not with the God of the Bible, because I do not believe the God of the Bible is the creator of this Universe and possibly others. My argument is with the logic people use, which in turn leads them to give their complete and total faith, time, and energy to the worship of these possible entities. This is usually done without any sort of thought or questioning, entirely forgoing critical thinking on these subjects for fear of deadly consequences by the hand of other people and the God they worship. Anything that revolves around fear cannot be of an infinite omnipotent faithful and unconditionally loving creator. When it comes to the phrase "unconditionally loving," there is no space in that phrase to be loathsome, fear-inspiring, and necro-mongering.

The fear in the religious text is extraordinary. The religious text has caused millions upon millions of murders since the time they were authored. Men, women, and children are being murdered all over the world presently as I write in the name of someone's God, and this saddens my heart deeply. I believe that if this God had a vested interest in loving and saving humanity, he would have done that a very long time ago, right in the Garden of Eden.

There is a reason I am diving so deeply into religion. Over the past few thousand years, nations of the Earth shaped their societies and laws according to religion. Up until recently, I would have been put to death for the slightest questioning of religious dogma. Even today, my examination could be met with anger and stir much controversy.

In the Bible, God has all the emotional and logical characteristics of a mere man. He expresses jealousy, rage, and favoritism, and even assists in war efforts. God does not put himself above any human expression, and nor does

Lucifer. So just who are these beings? Well, let us take a look at some texts written by the Sumerians believed to be one of the world's oldest civilizations. These Sumerians had myths that foreshadowed those of Genesis, and a numerical system based on the number 60, used in the clocks of today. They were usually given credit as being the creators of civilization being one of the first to use the potters' wheel and plow. But indeed, other civilizations predate this one, even some in South America that exhibit very advanced knowledge dealing with math and science, as can be seen in their architecture and what remains of their artifacts.

The main point behind mentioning the Sumerians is the fact that Eden is believed to have been in Mesopotamia. Other scholars place it in southern Iraq. Precisely where there has been untold carnage due to an American war. How convenient! Well, it is written in Sumerian cuneiform that the god Enki brought fresh water down to the Garden so that it may grow.

"HE then begat through the Earth mother goddess Ninhursag three generations of goddesses, all born with painless labor. Ninhursag, in turn, created eight precious plants that Enki ate. Angry, Ninhursag, declared that Enki must die and abandoned him. When eight parts of Enki's body began to fail, a clever fox persuaded Ninhursag to save the water god. Seated beside him, she brought into being a healing deity for each afflicted body part, one of which was a rib."(pg 16 Lost Civilizations Sumer Cities of Eden)

In Sumerian, the word for rib is "ti," so in turn the rib healing goddess was called "Ninti," which translates to two things, the lady of the rib, and the lady who makes live. In both Hebrew and Sumerian texts, man is made from clay. But a significant difference is that God ordered his mother Nammu to create man from the clay over the abyss. This is also very close to what the Gnostic text says. At least closer than anything I have read thus far.

Mother Sophia sends her daughter Zoë (Life) who is called Eve (Life) to raise up Adam. She sees Adam on Earth, an inert lump of mud, and pitying him, breathes spirit into his face. Adam rises immediately, and opening his eyes declares Eve the "mother of the living" for having given him life. (Manichaean Creation Myths (Gnostic); pg. 42 the Other Bible).

In the Gnostic Manichaean Myths, God had a mother named Sophia. A superior Mother God figure is also expressed in the Sumerian text. And God himself, according to this Gnostic sect who believes themselves to be the only true Christians, is the same as the Zoroastrian devil, Angra Mainyu, a secondary God below the mother Sophia and the father of all. They called this God of the Old Testament whom they believed to be malevolent, the Demiurge. They say that it was his evil act of creation, an attempt to trap spiritual light in ignorance and darkness, which brought misery and corruption. This is taken as being the true evil act rather than Adam and Eve's disobedience. The Manichaean Gnostics take Adam and Eve to be heroes for disobeying, who were aided by a Promethean figure, the Serpent, from whom they received knowledge. They believe that this figure later returned as Jesus to lead them in disobedience and to curse the Laws of Yahweh, the Creator. The fact that Christ, according to the Bible, redeemed us from the curse of the law is written (Gal. 3:13). The Gnostics not only wanted to free us from the laws but also the maker and the world he created, through methods such as extreme asceticism and meditation.

It seems as if the further back in time I go, the more unorthodox things become. How did Christianity and Judaism come up with their current canonical scripture when the origins of this scripture clearly state different accounts? It just merely depends on who you choose to believe when it comes to choosing your doctrine. However, the Sumerians do seem to be some of the first to give their account of this style of cosmogony.

So until further discovery or research on my part, I would have to say this is where our Biblical story stems. Everything else is just a copy of a copy of a copy. Even the Zoroastrian text is only a collection of older text.

I hope reading this is as much an adventure for you as it is for me because I am certainly having fun discovering new things as I do the research, and form new opinions as I write. First, I started with the cosmogony account of the Bible and voiced my opinions based on it, granted there are several days between each portion of the research and the writing down of new accounts from the Gnostics and the Sumerians. I express my disenchantment with the God of the Bible, whom I see as more of a mere malevolent entity than a benevolent one. It has not yet been revealed to me, that the Gnostic sect's view of our canon writ "God" as the Zoroastrian devil, Angra Mainyu. Though I am aware that they did view the canon writ biblical God as a Demiurge, I never guessed this view to be so strong as to give him the title "The Devil." This title is not of my creation, so do not kill the messenger. If the Gnostics view our canon writ, God, to be the evil wayward offspring of a feminine benevolent higher God, similar in respect to the Sumerian view of God, then how would they view our canon writ Lucifer?

According to the Manichaean Gnostics, Jesus was the Serpent in the beginning who told Adam and Eve to disobey God. This is just about the opposite of what we are shown in the canon writ Bible. According to the Bible, this act is committed by Lucifer. The Manichaean Gnostics ultimately turn many current views on its head. Calling the biblical God a demiurge, and the Serpent who is thought to be Lucifer "Jesus." So do I agree with these Manichaean Gnostics? Well, I will tell you what subject I vehemently do agree with. If there was a character named Jesus in existence, he didn't intend to make the allowing the desecration of his flesh, and his subsequent death, to be the primary symbolic example of sin forgiveness. But instead meant to show that this material world is a delusion that we are trapped in, and came to show the way by instructing us to dive deeply into inner light, and wisdom. But in Christian text, Jesus is a blood sacrifice; we will look deeper into that soon.

After all these points stated, I want to make it clear that I am not promoting a clear rejection of a belief in God. I am merely encouraging an educated questioning of our views with the support of research. I feel that this Universe has a prime source hence a creator. But does this creator have all of its morphological descriptions stated in the Bible? Is the cosmogony of the Universe written in religious text correct? I don't think so. I think many of the answers we are looking for can be found through traveling inward, and through meditation and research. Since we are part of the creation and creator, it is only natural that we can receive insight in many ways. I am amazed when I find out what I have come up with via contemplation, which at times matches what I later find written in other religious texts. Is this a mere coincidence, or do we have many of the answers, and are merely looking for validation? Either way, theological research is a fantastic way to gain insight into the world we are living in today and to educate yourself on what you think you already know!

CHAPTER 3

THE TOWER OF BABEL

Many readers of the Bible may be familiar with what happened at the Tower of Babylon. What's most interesting to me is why this happened. Why did God come down and confuse their language? They indeed could not have reached ontological heaven by building a physical tower. It's almost laughable today when you take a look at our skyscrapers, and space stations, and see that heaven is not located there.

Meanwhile, our ancient ancestors built a tower of great height to meet these places. So what was the true meaning of this story? How about we go over what is said in the Bible about the entire event? I will be quoting the New International Reader's Version, Which put things into a way of speaking that is a little easier to understand than the King James version.

("The Whole world had only one language. All People spoke it. They moved to the east and found a broad valley in Babylonia. There they settled down.

They said to each other, "Come. Let's make bricks and bake them well." They used bricks instead of stones. They used tar to hold the bricks together.

Then they said, "Come. Let's build a city for ourselves. Let's build a tower that reaches the sky. We'll make a name for ourselves. Then we won't be scattered over the face of the whole Earth."

But the Lord came down to see the city and the tower the people were building. The Lord said, "They are one people. And all of them speak the same language. That is why they can do this. Now they will be able to do anything. They plan to. Come. Let us go down and mix up their language. Then they will not understand each other."

So the Lord scattered them from there over the whole Earth. And they stopped building the city. The Lord mixed up the language of the whole world there. That's why the city was named Babel. From there the Lord scattered them over the face of the whole Earth.") (Genesis 11)

This portion of Biblical text is just loaded with mystery. It may be near impossible to investigate this portion of literature thoroughly and to come up with a foolproof explanation for every detail of the text. But there are some fascinating variables I would like to point out. For instance, the word Lord is used here quite often. What is a Lord? Here is Webster's dictionary definition.

Lord.1: one having authority or power over others: a ruler by hereditary right or preeminence to whom service and obedience *are due: one of whom a fee or estate is held in feudal tenure: an owner of land or other real property: the male head of a household: Husband: one that has achieved mastery or that exercises leadership or great power in some area <a drug ~ >: God: Jesus: a man of rank directly from the king: a British nobleman., etc.*

Lord is used very often in the Bible to refer to God. Why not just say, God? I believe it is because God also perfectly fits most of the many definitions of what a Lord is. He is one having authority and power over others. God appears to be male in all Biblical scenarios. And if you go to church, you would know that he charges a fee. Also, kings believe that they have been appointed by God to their position, therefore meeting the requirements that Lord entitles. The title of Lord gives them their right to rule over the land and the people who dwell on it. They also demand taxes, therefore doing God's work. So the Biblical God is indeed a Lord according to his purported will and behavior.

Along with the Lord, there are also others. We know this because he uses the word "us" as in the previous text. And again, we are left in the dark about who these others are. From what we have learned so far since the beginning of Genesis, it is clearly stated that the Lord does not work alone. There are always those others whose Identities are never revealed. But since Judaism is not the originator of much of the biblical text, we have to take a look at the writings from which it originates.

The stories from which much of our canonical script originates are documented to have come from a collection of stories from many sources in and around Egypt, as well as the near east. The Egyptians and Sumerians give a much clearer depiction of who the Lord may have been. They call them Enlil, Enki, Nammu, along with others. But in this case, I believe that Enki and Nammu are most pertinent. Enki, who has uncanny parallels to the Biblical God, ordered his mother Nammu to create man from the clay over the abyss. Nammu is the same as the mother Goddess Sophia in the Gnostic text and the Jewish pseudepigrapha. These Sumerian writings predate all known organized religions. The Sumerian Gods are known as the Annunaki. So what's referred to as God in the canon writ Bible may just be the Annunaki under a different name.

Our current civilization is modeled after the Sumerian and Egyptian civilizations. Our roots in the way we govern, educate, and entertain our people, according to modern historians, originate with the Sumerians. The Sumerians had no known predecessors to their way of living. In other words, no evolutionary links, which goes against modern historians' logic. Their civilization appears fully developed out of nowhere! According to Sumerian Cuneiform, they were taught directly by the Gods. The Egyptians also give many accounts of being taught directly by the Gods. Their civilization also came about during a similar period fully developed with no known evolutionary past.

I bet some of you are asking what exactly am I getting at here if it has not become clear to you yet. So now I will lay it on you. The God and the (us) of the Bible may not be any different than the Gods of the Egyptians and Sumerians of antiquity. According to the ancient text, these Gods gave us all the necessary knowledge and structure that mirrors our modern-day society, along with its major religions and governing structure. These Gods may very well be living entities. If this is the case, then these Gods/Elohim also created the world's major religions so that we may worship them under different names. And yet remain divided!

Let's further analyze the Biblical text. The people settled in this land now called Babylon. They said to each other, *"Come. Let us build a city for ourselves. Let's build a tower that reaches to the sky. We'll make a name for ourselves. Then we won't be scattered over the face of the whole Earth."(Genesis 11).* So here we see that these people who speak one language are united in a collective effort to build a city with a tower that reaches the heavens in the sky. They believe this monumental effort will prevent the division and scattering of the people over the face of the Earth. They also wanted to give themselves a common name.

If the goal of building the Tower of Babel was achieved without the interference of the Gods, I couldn't help but see this as anything other than a positive achievement. You have everyone speaking the same language working towards a common goal of their own free will to build a city and a nation where all the people of this Earth are united. There is no evidence of forced labor or slavery provided by the text. I am also quite sure that when they said they would build a tower to heaven, it was much more metaphorical rather than literal term simply meaning sky, and possibly a common heavenly state of mind—some utopia of sorts.

The Lord and whoever he was with, did not want this done, and would rather smite them with confusion and division.

"But the Lord came down to see the city and the tower the people were building. The Lord said, "They are one people. And all of them speak the same language. That is why they can do this. Now they will be able to do anything they plan to. Come. Let us go down and mix up their language. Then they will not understand each other."

So the Lord scattered them from there over the whole Earth. And they stopped building the city. The Lord mixed up the language of the whole world there. That's why the city was named Babel. From there the Lord scattered them over the face of the whole Earth." (Genesis 11)

My first question is, why did the Lord see this as a threat? This scenario draws many parallels back to the Garden of Eden. We previously explored the possibility of the suppression of knowledge enacted upon our human species for the sake of centralized control around a dominant entity. The scenario dealing with the tower of Babylon is not any different. Here we see the Lord destroying the unification of people who may have only gone light-years forward from this point in ever-growing knowledge and sophistication, in turn requiring less and less need for the control and direction of the "Lord." These people, or we, taking for granted this story, maybe based on historical facts, may have gone in a severely different direction from what we see currently today. Even if everything in the Bible was entirely fictitious, which I believe is unlikely, the stories told in the Bible would still hold the same amount of weight because the religious reader has faith that these words come directly from God. They require no evidence for this. Therefore these Biblical stories have the power of suggestion over religion-practicing readers. There goes knowledge... and with it the ability to control your destiny. Knowledge is indeed power. Take knowledge away from the people, and with it, you have taken away their power.

I know that I may be upsetting many readers out there. I may be doing something precisely akin to blaspheming. A couple of hundred years ago, I may have been burned alive for saying these things. I don't doubt that some of you may want to burn me alive today. But that would not be necessary. I have been told never to question the ways of the Lord, and the answers to the things I didn't understand about God would be that he works in mysterious ways, and I was advised to quickly sideline further questioning for fear of the Lord, and the fear of his wrath upon me. I was raised as a Christian in a Christian household. During my teen years, I carried unshakeable guilt, because I knew that I could never live the perfectly sinless life I was taught I should be living, and I knew that no one could. It is virtually impossible while still living in a human body.

My mind never stopped questioning my surroundings as a teen, and the questions were only getting bigger and deeper as I mature. There was a part of me that knew one day I would reject many of the ideas pushed on me by my Christian faith, because of the conflicting knowledge I would discover via my questioning. My mind was thirstier than the driest of deserts, the emptiest of voids, like a low-pressure system sucking in air and matter from fields of higher pressure and greater density. I was not satiated by the things I learned in school and church. All the while, believing that these institutions were becoming increasingly useless when it came to the answers I required in this life. I had complete and total awareness of the dissolving indoctrination forced upon me by today's standing institutions. I would daily ask God if he is real, and if he is who the Bible says he is, then to kill me and take me to his heaven. I knew that sooner or later, I would stray from his perfect requirements and turn hell-bound, and I feared nothing more than my soul burning for an eternity in a lake of fire. I would specifically ask him to kill me, but painlessly. I didn't care about dying as much as the pain that might come with it. That's all I would ask, and to do it during a time when I had a clean slate, to prevent myself from burning for an eternity before getting a chance to beg for the forgiveness of my sins.

In other words, I lived in the dark because my mind did not have specific knowledge. Powerless because I was dictated the truth by institutions of particular interest. And Last but not least, I lived in fear because I believed that the Lord of the Bible would be upset with me, and would partake in my eternal burning in a lake of fire. A neverending burning offering! What a fantastic appetite! For anyone or anything.

I am sure that after the publication of this material, while living a life surrounded by religious zealots, many people are going to be highly upset with me. That includes both of my parents, relatives, friends and many others not related to me. But I say to you if you love me, then why does it matter what I think? Get out there, think for yourselves, and do your research.

Isn't it interesting that many things seem to be holographic in this Universe? Figuratively and literally. Our minds still work the same way as our ancient ancestors. That is why my emotional state of mind could be mirrored by Biblical characters and archetypes. Thus making their specified scenarios relevant. Thus placing a cap on my intellectual development for fear of supernatural consequences, along with the fear of parental and societal backlash. Luckily I was aware of it and dared to seek answers. But many people are not so lucky and will live the rest of their lives in the cage they have built for themselves in their minds.

Once again, I would like to reflect upon Biblical scripture. The Lord confused the language of the people so that they could not understand each other and scattered them across the globe. Now for a moment, take a look at a map. You will see that there are many nations with most of them speaking their unique language, and these nations are scattered across the whole globe, many of them still warring against each other to this day.

Is it possible that this problem could have been solved millennia ago? I think it is possible. According to today's scientists, all people on this planet share a common ancestor and have migrated from the same place, which is the general region of eastern Africa. Our genes say so. This area on the globe is not far from Babylon, a region easily accessible to early migrants. So now, let's check our common sense. Doesn't it appear logical that since we all came from the same people, we might have all spoken the same language at some point in our past? Think to yourselves about that.

Let's remember the main reason for the Lord scattering these people, (our ancient ancestors), was to prevent this force of unified people from doing whatever they set their minds to achieve. *"They are one people. And all of them speak the same language. That is why they can do this. Now they will be able to do anything they plan to. Come. Let us go down and mix up their language. Then they will not understand each other." (Genesis 11)* So I am sure you are asking yourselves why the Lord would want disunity amongst the people, and would rather have them live below their full potential. Well, if you simply look around and observe the current state of our world. You would see that we are in a constant state of war to divide and conquer unified people to assert control over them easily. No other reason exists, other than for the selfish gain of the dominant controlling power. The Universe is holographic. Mirrors are everywhere.

CHAPTER 4

SACRIFICE TO THE GODS

Sacrifice or offerings to a god or deity is an ancient religious practice. The reasons for making a sacrifice or an offering can vary depending on the religion, or the criteria the entity requires its subjects to adhere to. Yahweh, the God of the Bible, also requires a sacrifice or a burnt offering. The first one recognized by the Bible is the incident involving Cain and Abel.

Adam knew his wife Eve intimately, and she conceived and gave birth to Cain. She said,

"I have had a male child with the Lord's help." Then she also gave birth to his brother Abel. Now Abel became a shepherd of a flock, but Cain cultivated the land. In the course of time Cain presented some of the land's produce as an offering to the LORD. And Abel also presented [an offering] — some of the firstborn of his flock and their fat portions. The Lord had regard for Abel and his offering, but He did not have regard for Cain and his offering. Cain was furious, and he was downcast.

Then the LORD said to Cain, "Why are you furious? And why are you downcast? If you do right, won't you be accepted? But if you do not do right, sin is crouching at the door. Its desire is for you, but you must master it." Cain said to his brother Abel, "Let's go out to the field." And while they were in the field, Cain attacked his brother Abel and killed him. (– Genesis 4:1-8)

The first major question I am asking myself is why does God require an offering? What purpose could burnt plant matter or burnt animal flesh serve to God? This has never been clear to me. The Bible gives no clear explanation as to why these things please God in any way. What we do know is what the Bible tells us. Abel is a shepherd of a flock of sheep, and Cain is a cultivator of land. Abel gives God a burnt offering of the firstborn male sheep, and Cain being a cultivator of land, offers God some of the things he had grown. From my perspective, this sounds very logical that one brother would give God a plant offering and the other a living creature, being that one brother was a cultivator of land and the other a shepherd. During this period, these two occupations hold an equal amount of importance. But God, for some reason, does not accept the offering from Cain, God seems to prefer an offering of burnt flesh, over one of plant matter. So what does this tell us about God?

Well, I see some symbolic significance in this story. Abel is a leader of a flock of sheep; the firstborn males to female sheep were classified as the best offerings. So for the sake of gaining a different perspective on this story, let's say these sheep were people. So Abel is now a shepherd offering the "people" to God. The firstborn sons of the mothers, and the fattest parts, the best of the best amongst the people. Being that Abel was a shepherd, this would make him also a leader. Whether it was a flock of sheep or a multitude of people, the archetypes of this story are pretty much interchangeable. When it comes to symbolic meaning, a flock of sheep can easily mean a crowd of people, hence the slang word sheeple. While Cain offers God some plants which he has cultivated. God has no interest in a vegetarian diet. God wants the flesh, be it animal or human.

If you guys don't believe they were other humans besides Cain, Abel, Adam, and Eve, then where did Cain get his wife? It certainly wasn't Eve, and while getting his wife pregnant, he proceeded to build a city! So that means there is an awful lot the Bible isn't telling us, and that there had to have been countless thousands of other people living during this time to occupy a city.

So let's say we take this story at face value just as millions of people around the world are instructed to do. When God accepts Abel's offering, what does he do with it? Does he eat it? Or does he store it away for safekeeping? The Bible doesn't say. If you are hesitant in believing that God would not accept a Human being as a burnt offering, take a look at the story of Abraham and his son (Though God stopped this sacrifice short). Humans are food for the Gods spiritually and possibly literally.

The Sacrifice theme is found in cultures all over the world, and the Bible is no stranger to it. Off the top of my head, I can think of about two other instances of apparent blatant sacrifice in the Bible. In Exodus, the killing of the firstborn sons of Egypt during Passover, and Jesus in the New Testament becomes the ultimate sacrificial lamb. The ender of all burnt offerings. Something like Osiris.

Man was made to sacrifice to appease God or gods from the very beginning, hence (Homo Necans) "Man the Killer." Ptah was Egypt's earliest civilizing God, who also brought fire, but also ritualized bloodshed.

Inscribed on the walls of Pharaoh Wenis's pyramid between 2375 b.c. and 2345 b.c. is written an incantation called the "cannibal hymn" Wenis is described as "he who eats men, feeds on gods." "Devine hawk" who "devours whole" those he finds on his way."

Osiris, the great-great-grandson of Ptah, who ended the practice of human sacrifice, ending the curse of Cain, is remembered with warmth and gratitude and is Egypt's greatest civilizer. Osiris was also the first to make humanity give up cannibalism. Cannibalism is one of the oldest and most fundamental forms of human sacrifice. Civilizations such as the Aztecs, Greeks, and many others are known to have cannibalized. Like the later Aztecs, Egypt was a civilization built on blood up until its first dynasty, 3150-2925 bc. After the cult of Osiris came into power, ritualized sacrifice went into sharp decline.

The reason why I bring up Egypt with the mention of Cain is to bring forth the fact that the text in the Bible is thought to be Hebrew in origin by people who practice the three major religions, such as Judaism, Christianity, and Islam. But the sources for these stories are Egyptian and Sumerian, amongst other neighboring nations in that region. (Same stories different names). Egypt also has its fair share of Christ-like figures.

All around the world during the time of the ancients, human beings were sacrificed and buried beneath new buildings. This is an ancient practice also found in Hindu scripture. In German folklore, children were buried alive, which still may be practiced by the Indians in the Andean communities to this day. These are called "Foundation sacrifices." Foundation sacrifices can be found in the book of Enoch to convince the gods to allow the buildings to stand, a type of bloodletting to appease the gods or God. Victims may haunt the structure forever, serving as spiritual guardians. According to the Bible, when king Bethel rebuilt the city of Jericho during the reign of King Ahab, he sacrificed his firstborn son Abiram for the foundation and his youngest son Sequb for the set up of the gates. This is known to be a Canaanite ritual often found by archeologists commonly in Syria and Palestine, where infants are found beneath foundations. The rituals were done to strengthen the structure being built. This practice may have been carried on in tradition by "freemasons." The freemasons may hold on to the culture of sacrifice through a form of bloodletting we know as war.

In Egypt, these practices were replaced by nonhuman sacrifice as the result of Osiris by using a substitute sacrifice, or a shabti, such as bread, wine, oxen, or even a small carved figure. Osiris is believed to have marched over every country in the Mediterranean, teaching a gentler manner of a non-sacrificially centered civilization. Osiris taught winemaking, farming, and metalwork, and was worshipped as a god to the dawn of Christian times.

Many ancient myths that led to the stories of today are told in such a way that they seem to have a common origin. Cultures around the world separated by thousands of miles go through the same ceremonies and traditions. All the gods and saviors of the past have remarkably similar descriptions {Vericocha of Meso America, Horus, Osiris of the Egyptians, Jesus of the Nazarenes, along with many others not named here}. It seems that in many cases, they are the same gods with different names. The same with today's religions, I call him God, you call him Yahweh, they call him Jehovah, and others call him Allah. In a nutshell, what we understand as religious truth, is a very ancient myth or story passed down through time and changed to fit whatever the leaders wanted to get out of that specific culture or nation during that time. There seems to be a pattern here. As a Disclaimer, I am not saying that these gods did not exist in any way.

In the Christian religion, Jesus is the Sacrificial Lamb, having died for the atonement of our sins so that God could finally accept his wretched human creation. In the Christian church, we engage in ritualized sacrifice and cannibalism of Christ through something called the "Eucharist." I believe this is done on the first of every Sunday of every month in most churches. To participate in this ritual, you must symbolically eat the flesh and drink the blood of Jesus. This is done by eating a cracker or wafer, which represents the flesh of Christ, and then following up with drinking wine, or some sort of fruit juice which represents the blood of Christ. This is very similar to the shabti concept coined by the Egyptians (Ptah, or a substitution sacrifice. I believe this is nothing other than an ancient tradition under a different name, the cannibalism of christ

I would like to voice my disclaimer. I am not an atheist or a hater of God or religion. I only hate the ignorance that it is steeped. It seems as if most people leave all the questions and research to someone else while allowing themselves to be led along in blind faith. I wholeheartedly love the creator of this Universe and perhaps multiple universes and all things in them. Religion, which is driven by what seems to be more of a personal agenda rather than love, could not possibly be describing the heart of an infinite creator basking in the pleasures of all that is. All religion, in my personal opinion, reflects far more the heart of a mere man, or an entity infinitely lower, than our all-encompassing creator.

When it comes to how I stand on Christ, I come to find out that my views reflect those of the Gnostic Christian sect much more closely. They believe that salvation is not up, or down, nor high, or low, neither here nor there, but from deep within your self. You are the only one who has the power to change yourself and save yourself. The Gnostics believe that Jesus' or Christ's conscience was trying to show humanity that the world you see is a mere delusion or illusion and that the light of God was already in every one of us. Due to this truth, we all can heal and transcend this illusion we call reality. Some believe that Christ's primary mission on the cross was not to be seen as a sacrifice and to be worshipped as a God. But instead, it shows that the body and the world around us are an illusion. The body is merely a vessel, and he did not suffer, only because he understood the nature of this reality. This is the ability that Christ believed that we all have, the ability to see the light within and transcend this world of illusions to save ourselves, through love, and inward change through "knowledge." Hence Latin word Gnostic, which means: someone who seeks knowledge. That is the gift of Christ.

What is the act of worship? First, let's take a look at its definition and root meaning.

Worship: worthiness, respect, reverence paid to a divine being or supernatural power. Extravagant respect or admiration for or devotion to an object of esteem.

The root meaning of this term: ***Worship*** *is etymologically derived from* <u>*Old English*</u> *words meaning "worth-ship." Giving* *worth to something. In its older sense in English of worthiness or respect (Anglo-Saxon, worthscripe), worship may on occasion refer to an attitude towards someone of immensely elevated social status, such as a lord or a monarch, or, more loosely, towards an individual, such as a hero or one's lover.* (according to Wikipedia)

When you worship something or someone, you are focusing your energy on the object of worship. Why does the God of known major religions require that humanity focus their collective energy on him? What is the purpose of this, and how would this action benefit a supposedly infinite and perfect creator? What could the worship by beings on a planet that is so small that it couldn't even call itself a speck of dust when compared with the size of the Universe, do for an infinite creator? I just don't know. I could not even wrap my little mind around it. But maybe it does something. Perhaps something just beyond my range of understanding in these moments. But there is one thing I am sure of. When you worship, you are giving away your power to something outside of you.

So now that we are complete with our general discussion on the Lack of basic religious understanding most of humanity shares, what have we learned? Personally, what I have learned is never to be afraid to question the status quo, because where there are questions, there is a general lack of knowledge, and the answers should be sought. "Walk in blind faith alone, and you will be walking the path where the blind do sleep. Open your eyes and see, and the light of knowledge will wake thee." (Marcus Morris)

CHAPTER 5

THE BABIES FIRST CRY

Why did the first baby let out its yearning cry? I think it is apparent to any mother that the baby lacks something, and the mother must figure it out. I believe that it is reasonably evident that the human adult is fundamentally no different than what it was in its infant past. What has changed is simply the scale and complexity of the person's needs. Instead of being satisfied by the mother's milk, that same individual may feel the need to indulge in a lobster dinner. While an infant, you may want to suck on a pacifier, upon growing up, require a beer or some wine to fill the same niche (self-soothing). An infant seeks comfort and happiness, and as an adult, you continue to find comfort and satisfaction. Adults are nothing but big babies. The only thing that has changed here is the scale and complexity of their being and needs.

From birth, we experience a Lack of the coming of our material death, and in death, we lack life. What I have noticed here is that the word (feelings) becomes super important.

More important than anything else I could wrap my mind around, as long as you exist, you will feel something. No one is exempt from this. The Universe is a realm where all sensory information brings about reactions we call feelings. We feel hungry, sad, anxious, happy, and afraid; all things revolve around the sense. There is no experience without feeling. If you disagree, go and test yourself. Try not to feel or to feel nothing. Is it possible? I don't think so.

The concept of Lack and the act of feeling are inseparable.

There is so much you can learn from yourself when you listen to the emotions that are speaking to you from your emotional center. This is where all the answers to who you are and what you want, exist. So what is the major problem here with humanity? The feeling of Lack is the problem. Lack is a symptom of separation or a void that sums up a whole plethora of things. Separation is the empty void between us and whatever is it that we lack.

If any Human being wishes to understand themselves completely, they must be willing to start from the beginning and examine every memory that has played a part in molding them into an adult. In other words, start from birth. Since most of us cannot remember that far back, start from our first memories and enjoy them, that is why they are there. But do not skip over the bad memories, because here is where the lessons reside and the keys to the questions we ask about our selves. In these bad memories, lay all of our fears, pain, anguish, and insecurities. Unless all of these memories are dealt with correctly, all of your negative thoughts and feelings will still exist, there is no covering them up and suppressing them as long as you still remember them. So in healing, you need to make a bridge through the separation you feel to that part of yourself that has been damaged and rescue it. Simply love that part of yourself and accept it as a part of you. This is a great courage builder for that person who may be trying to achieve great things in life, or simply anyone in general.

The usual result of the damage done to the individual is that they become a shadow of their former self. A sort of dissociation occurs within the mind, and the Ego begins to take control as a form of protection. The Ego usually takes precedent in most individuals and overdoes its job of protecting the individual. It becomes a mirror of whatever is outside of you. It has been working throughout your life to mold you into who you are today. But this mirror also goes both ways. What's inside you is also being projected to the outside world (thoughts, feelings, etc.). It exists in the form of negative or positive energy. Its something like being the police force that is continuously trying to keep yourself in check, for fear of what others may think, in turn hindering your individuality. Now imagine this on autopilot. That is the relationship between your mind and its Ego over a lifetime. Your mind gets thrown out of balance because the Ego usually gains the upper hand. In turn, you spend the rest of your life seeking balance and primarily yourself. But remember, yourself is never far away; simply go inside and open your heart and listen to your inner voice. You have all that you need right inside you to feel complete and whole again.

The evidence for my previous statement lies within the very beginnings of your bodily existence. We tend to call this childhood. Did you not feel whole, complete, and loved, only as a state of being during some time in your life? Sometime back before your inevitable conditioning to your current state of mind? I bet you did; I think it is safe to say that most people enjoyed this state of mind when they had it. Since we already know that this state of mind is a natural part of our being, there is no reason why we can't have it back and keep it forever, right?

Children still have their innocence because they have not gained the knowledge of good and evil, and begin to develop feelings of guilt and self-conviction. After this is done, the innocence is gone. They have tasted the proverbial apple designed by the society we inhabit. I couldn't say whether this first moment is good or bad, but it usually steers out of control in any given individual when the Ego takes over. Learning has been turned into an arduous process in our western society. Learning is done with the Ego, and in our western culture is designed with the intent to have the Ego take complete and total control of the individual, in turn alienating your true self.

I disagree with the cut-and-dry teaching methods today's institutions promote. Learning can be done so that it is a nonlabor-driven experience, and instead designed to capture the joy, and wonderment of the experience of life.

I suspect that original sin is moving away from our original state of being, which is wholly innocent and in love, compared to our present conditioned state ruled by the Ego. Human nature governed by Ego is a constant battle raging within. Our world reflects this, which is filled with wars and countless issues. However, there is a lot of good in this world. What is inside is reflected like a mirror onto what we experience as the world outside. There is more than enough beauty and real wonders to inspire humanity to change for the better. But it all starts on the inside because whatever happens inside is reflected on the outside. So focus on the wonderment and beauty we all inherited as a human species, the brothers and sisters we all are. It's as simple as changing yourself. "I'm starting with the man in the Mirror." Michael Jackson"

The baby's first cry is synonymous with feelings of Lack, and these feelings only increase and magnify from this point on as we begin to realize what is out there that we don't have, and want to acquire. But upon becoming self-aware, which is merely being aware of the thoughts racing through our minds on a day to day minute by minute basis, you can begin to control your emotions and desires, instead of them taking control of you. This is not a control where you overpower one feeling with another. The control I'm speaking of is done through wisdom and awareness. You can master your emotions when you know them and understand why they are there, and what they are teaching you. Once you gain this understanding, you can learn from your current emotional state and move on naturally to whatever is next. It has been done for thousands of years, and they have called this inward (thought travel) Meditation. "You do not have to be an expert or a yogi to do this, and it is not dark magic or witchcraft. It's simply getting to know yourself on a more intimate level and becoming the master of oneself. Don't you desire to be your own master? I'm sure that most people do. Even the blue genie of "Aladdin" desired to be his own master, and I certainly do intend to master myself my most high potential.

Many people may look at this as perfecting oneself. But if you are looking to be a perfectionist, know that an -accomplished perfectionist does not look for imperfections any longer to improve, but rather, looks over the flaw. They instead, acknowledge what is perfect and then settle. Therefore you will remain stagnant in certain areas of your life. Perfection has no clear limits, so simply enjoy and improve instead of seeking perfection.

CHAPTER 6

THE DREAM

Reality can be described in such a way that one is within oneself dreaming. The brain is a decoding processor. The brain receives abstract symbols and codes which come in the form of sensory stimuli, then crunches down the information into organized mathematic symbols or forms of energy your brain can recognize. All brains work this way. Anything outside of your mind that you classify as something to help you think or solve a problem is a tool. All devices fall under the category of things decoded by the mind, to create symbols we understand.

The main difference between this present state of reality, and the state of existence when we drift into during R.E.M. sleep, is that when you are dreaming, your world is simply projected inward. So instead of relying on your eyes to experience reality, the inner workings of your mind become the dominant force putting to use an organ called the pineal gland. The pineal gland is a small organ at the center of your brain covered with the same rods and cones within your eyes. Like your eyes, it is also filled with fluid. The pineal gland operates like an eye that has no direct contact with the outside world and instead controls all interactions with the inner world. It is what the ancients called the third eye. The third eye is depicted in ancient art as something that looks like a pine cone across many cultures.

The Pineal gland has also been called the God gland or a stargate; it is believed that the pineal gland connects you to the whole of the Universe through a Unified Field. Upon opening your third eye "pineal gland," it is believed that you could achieve all sorts of extrasensory abilities such as telepathy, and clairvoyance, through practice. Based on the metaphysical nature of things, we are inside ourselves dreaming, or in other words experiencing ourselves through the use of symbolic archetypes to evoke an emotional feeling response. We do this collectively. The reason why we can do this together is that separation is an illusion, and we are, in fact, one.

In this dream, we live through many illusions that we make real when they are merely thought symbolic archetypes. The fact that we are convinced of the realities in which we live reveals how powerful they are. If they weren't as powerful as they are, we would all be aware of the illusion, and the belief in the Universe simply wouldn't work out.

I once heard an astronomer give the most compelling hypothesis on the nature of the Universe. The man stated that there is no beginning or end to the Universe The Universe instead recycles itself forever. Similar to the illusion that you see outside of a barbershop with the white and red cylinder spinning. The Astronomer also stated that if you took a telescope and tried to peer as far as you could into space, you would eventually be looking at the back of your very own head. I thought that these were some of the most provocative statements I have ever heard. I wish I remembered his name because I certainly would like to give him credit for these theories. Somehow, what he says may fit in with the metaphysical nature of this Universe concerning its relationship with thought and mind. This statement is so important because we are one consciousness and really one conscious mind. This mind is seeing its self from infinite scenarios and perceptions. So as we search more in-depth into the Universe for its beginning and end, we are still that conscious, biological self-aware mind of the Universe looking at ourselves because we are the Universe. Matter cannot be created or destroyed; it only changes state. (Law Of Conservation Of Mass)

I have a theory of my own. The Hubble space telescope can perceive something they call redshift. Redshift happens when objects are moving away from us. The farther away things are, the more significant the redshift, which means the faster the object is moving away from us. So when we start to look for objects further than 13-14 billion light-years away, which astronomers claim to be the age of the Universe, I believe that we can not see these objects because they start to approach the speed of light at that distance. Light from those objects can no longer reach us because it is now traveling at the speed of light away from us in the opposite direction. I hypothesize that the Universe is not 13 billion years old but instead infinite. Then again, all perceptions of things around me only occur in my mind.

Change your mind you change your reality. All things of your existence and what you think is real in this world are located somewhere in your mind as an electric pulse, with the change of mind, a shift in perception in reality follows. As a result, I realized that stress is a mutable state, and I have a choice. Especially since stress is nowhere but in my mind. Can I find stress outside of my thoughts? I admittedly haven't seen it yet. But I will keep looking.

All fear also has its roots. Fear stems from the belief that you will someday die along with physical and mental pain. However, death is the most prominent of these three. We all have a limited amount of time here on this Earth. Knowing this, we all must abide by the dimension of time. Without the belief in a limited time, fear would be virtually obsolete, because we would have an eternity to do and create anything we wanted in our lives. The unknown of what happens after we die keeps us on our toes. For example, I feel that I have a limited amount of time to write this book for many reasons.

Contrary to the belief by many religions that God is to be feared, to me, it is more apparent and logical that the higher your fear, the further away you are from God and Love. Lose your fear, and you can stare directly into the eyes of God, the creator of all things great and small. In my opinion, Fear and God do not mix. This is a significant reason why I do not accept religion as it is because religion is such an immense bearer of fear and uses fear as a vice to subdue its followers.

The sort of fear used by religious leaders has a potent effect on its subjects because it deals with one of our great unknowns. Which is, what is going to happen to us after we die? Religions take full advantage of human beings having no idea who they are, why they exist, and what is going to happen to them after they die. Religions claim to have all the answers to these questions. In return for religious reassurance, you must give full devotion and self to these Gods. Walk on eggshells for the rest of our lives, natural lives for fear of sinning. Sometimes this means going against our instincts just so that we can spare ourselves from an eternity of being tortured in a lake of fire. All questioning of religious methods is called blasphemy. And if you are near perfect in your devotion, you will still fall short of glory. What a dilemma! This is why I answer to love and only love, the essence of the actual creator. I believe that if God, the creator of all things, wanted anything from us, it would be our ever-increasing awareness of the totality of existence, ourselves, and Love.

Pain is there as a prevention method to keep us from falling into death. You know that when you have pain, you instinctively understand that you need to do something to stop the pain. Not too much of a mystery here. But what is interesting to me is that the higher the magnitude of pain, the more we fear it. From my observations, the greatest fear of almost any human and any other living creature is to be eaten alive. This simply seems to be a fear that is most prevalent in dealing with the animal kingdom. The other known fears, such as falling from a great height, drowning, burning to death, etc., are pain-related also.

In the whole scheme of things, what is the archetypical symbolic significance of the entire fear-pain connection? What is it that our souls are supposed to learn from these things? From my observations, I see life with all of its complexities as an excellent learning tool for the soul, if you have built up your awareness enough to look at everything from that perspective. So what has become apparent to me is no matter how far you go into the deep dark corners of life and all the challenges it throws at you, the power of the mind is always present, which is like your central computation station. With awareness and presence of self, you can use absolutely anything to increase your internal power and essence and come to understand that you could never truly be harmed nor broken. We are eternal consciousness. Thus gaining the ability to manipulate all thoughts going through our minds to determine the best possible scenario for ourselves

In turn, you realize the puzzle and take things into accord while loving and living in the essence of the eternal self. Fear and all other negative thoughts dissolve away upon one's own will. Pain and death are temporary. The spirit, the soul, and consciousness are forever.

What these thoughts led to is another realization. Once you start the momentum of self-awareness, it just doesn't stop, and a whole new paradigm is opened up to you.

What I realized is that we are not our actions. Society has conditioned humanity to believe that we are our actions. For instance, if you ask someone what they do for a living, they would describe their jobs or professions as what they are. Such as, someone would say that I am a lawyer, or I am a doctor, etc. But these things are not you. They are your actions; actions can not be you. Not only does this cover what you do for a living, but also in describing your religion, sexual preference, political affiliation, etc. What this conditioning has done is make you believe that you are your action so that you are trapped in that mode of thinking. This mode of thinking makes you think that you are separate from significant components of yourself, and from others who may think differently than you. It is one of the most awesome divide-and-conquer tools. Even Race falls under this category because people feel that they must conform their behavior to the race they have been socially forced to divide into from birth, in order not to offend other people and the status quo.

People are also taught that they are sinners from birth by religion. We are told that we fall short of the glory of God. It looks quite apparent to me that religion teaches us that we are our actions and that God also believes this, and will judge you according to your behavior or actions. I think that the whole idea of assuming someone is a guilty sinner from birth is just a bad idea for the mind and soul going through a life experience here on earth.

The whole practice of assuming guilt within oneself could lead to committing more evil behavior, which is more in line with the purpose of guilt conditioning in the first place. In other words, the person who already has a guilty heart is more likely to undo themselves to mold themselves into that specific pattern of deprecating behavior because it is dominant in their mind. What you think about you bring about. The same person who has innocence in their heart, and thoughts, and believed they are born on this planet clean and innocent, that same person is more likely to lead a life of benevolence and innocence because those are the dominant thoughts going through the mind of that person. "Thought always precedes action." So if your thoughts are filled with guilt and regret, that is what you will get out of life. So believe in your natural-born innocence, and your life will reflect that.

The Media works hard to perpetuate the guilt and ill will of humankind; this is obvious because the news is more than 90% negative. So it leaves the conscious and subconscious psyche feeling downtrodden.

The dream is all in the mind, along with the universe and all other points of perception. Wake up and stop dreaming. Reboot, become aware, and then "wake up and start dreaming," and your awareness and reality will be changed forever.

CHAPTER 7

FUNDAMENTAL EXISTENCE

From my observations, it seems to be true that we all are essentially part of our sun. Scientists have said that we are indeed made of stars, but I want to take this truth a bit further. All objects, this includes the planets and their moons and all other bodies that revolve around the sun since its inception, are more like cool matter in its outer atmosphere rather than separate entities unto themselves. This would include all living things native to this solar system.

I think that it can be proven sooner or later that all matter is frozen in its present state through a type of sound or resonant frequency, hence music. If it has not been proven already…! With that being stated, it would hold that no object or living thing can change its form or become a new species unless its resonant frequency is changed. So this would put a unifying force in charge of all things that exist, which controls things by tuning resonance. Natural selection on its own can mean whatever you or any scientist projects onto it. Natural selection by itself cannot explain things on multiple levels. The Fibonacci sequence is seen in all living and non-living things. The Fibonacci sequence is also recognized in the spin of a galaxy or the swirl of a hurricane. These phenomena need an explanation.

Resonance is more of a very complex language, which is the template for creation. Some species have physical characteristics that are designed for a symbiotic relationship. Meaning one plant or animal species is designed for only one other plant or animal species on this entire planet, like parts in an engine that makes it function. A Moth in Madagascar has a foot-long proboscis made for only one type of flower. Each of them would die without the other. As if they were made simultaneously one for the other despite them being TWO very different unrelated species. There are many examples of this in nature, and evolution based on natural selection alone cannot logically explain such occurrences.

Frequencies may play our DNA, which is also made up of frequency like an instrument to manifest different physical characteristics. So when the Universe or the Sun changes the frequency it emits (most relevant to the sun because it is our closest star), we change /our DNA changes. Being that we are part of the Universe and not in any way separate, these thoughts feel very intuitive and logical, so they hold a great truth for me. It seems apparent to me that anything that receives heat or any form of energy from the Sun is part of its outer atmosphere. In part, this would make us Human beings, intelligent Sun Light forms. Self-aware Sunlight! Take it a step further, and we are self-aware Universal starlight.

The Universe is resonating in unison on some level. This is why the universe works, and things fit together in a synchronized way. I could easily imagine the Universe having no unity through resonance, and what I see is a universe that does not function. The forces that act upon all living and non-living things could not hold atoms together to form matter to then form stars and planets without a base resonate frequency. If you take a look at an atom and then look at a solar system, there are stark similarities in the design of these two systems. So without resonance, an atom would not be recognizable. I believe it would disperse into nothingness because the forces that act on it would be missing, so no form as we know it could be held.

Then again, all views of the Universe only occur in the mind and not somewhere out there. Anything observed with the senses is happening in the brain. So all things involving science, and all other experiences stem from human perception. It all occurs in the human mind, and we cannot imagine what it is like to perceive anything from a non-human perspective. If you are human, it is impossible to do so. Just like you cannot imagine non-thought, which cannot be affected by thought. This is why all thoughts are equal because they all are dealing with the human perspective, which may not give an accurate depiction of the nature of all things known. But then again, it may be because as we are observing, we are gaining an understanding of our mind from our human perspective, which is a reflection of the universe we perceive as being outside of us.

Imagine that everything that you thought was instead just a trick of the mind, the power of the mind to make things seem as real as they are perceived. Imagine that a pencil might not be a pencil at all—instead, just a blob of something inconceivable. Our perception is filtered through our human senses. Which means there are likely infinite possibilities outside of our human senses. I think this indeed may be the case when it comes to our material existence.

Let us take this a step further. One cannot understand or comprehend anything that they do not have receptors for, hence (reception), for a frequency, like a television. You must have some sort of receptor for decoding information to perceive the subject. Without this ability, the information will be non-existent to you. Try plugging an American appliance into a European outlet. It doesn't work. You need a converter/ a modifier/ bridge/ new receptor/. The mind works this way; the universe works this way. In a nutshell, just because they may be things in this Universe that you may not be unaware of or do not experience regularly, does not mean they do not exist.

In retrospect, nothing can indeed happen anywhere else besides the mind, down to every last detail of your observable life. If you are a good conductor of your mind, alluding to the way a conductor commands an orchestra, you can get your thoughts to play any symphony of your choice.

The Universe is a reflection of our imagination. Your thoughts and memories make up the architecture of your universe. Nothing that you do not know can be in your Universe. Those things can only exist upon discovery or making it up via imagination.

"I/ You" am the singularity that experiences all things outside and inside of myself. Separation is never actual. My hands are my eyes, just as my ears are my eyes or part of the same ultimate tool of sensory perception linked to my thought, which is whole in itself. Inner exploration is the key to the one real source of all experience.

Because of what I observed about human nature, I give us a new name. I call us "Homo Measuresus Judgementus Adjustus." The explanation is in the title, humans as a whole measure, then judge, and adjust to whatever the scenario demands. The measuring is the analysis of the information when taking it in and figuring out what kind of weight it holds on our existence and well-being. Then we judge the data and adjust our behavior accordingly. The reason why I put an "Us" at the end of the words is that it deals with our relationship with each other and how we read cues from one another and the environment. We measure ourselves, judge, then adjust ourselves so that we may deal with whatever is at hand. This is all done automatically or without conscience awareness with most people. Only when you become self-aware will you start to observe your thoughts.

Every person goes about their entire life living out the "Homo- (meaning same) Judgementus Adjustus" experience. The way this is done in the mind is mathematic in a sense, but not the digital type taught in schools, but more "analogous" such as an old fashion film camera. I see this as being more in line with the intuitive portion of being.

Speech, for example, is nothing but our bodies creating mere vibrations through the air that the receptors we call our eardrums pick up and is decoded in our brains into something we can understand. This is our bodies going through measuring judging and adjusting processes, so then we hear things we comprehend. This is done on many levels dealing with the processing of information.

I want to throw this bone in there. "Normal is only the average state of mind of a collective of weird people." I thought this self-quotation was quite funny and relevant to this collection of information presented here.

The way I see it, this whole (Muniverse) or mind-verse, is nothing but a joke to be laughed at. Everything is thought, and things may manifest according to what your mind is focused on. It would be much easier to live this way in practice if things didn't seem quite so solid and real. Maybe if matter had a vibration that was a little more malleable, then I would have an easier time putting these thoughts into a proper perspective.

Nothing thought generated can be real, only because nothing thought produced is permanent. All thoughts are fleeting; this is why we are not our thought or actions, and neither one of these are permanent. We could only be non-thought, the stillness and permanence. If you tried to think about this stillness and stability, you would fail because you would be thinking, and by now, we know that all thoughts all fleeting. So it would have to be something that is felt. More like the feeling of being alive, and feelings of unconditional unshakeable love. (Which is something that is probably not experienced here on Earth very often). Most relationships come with there silent agreements on behavior that each participant must obey, even in successful relationships. If the statutes are broken, the loving relationship will fall apart. I'm not saying that most human beings are unable to express unconditional love; it's just that a lifetime of fear conditioning, stress, and individuality-suppressing dogma build mental blocks that prevent access to this part of yourself.

Unconditional, everlasting love is real; joy and laughter are real. So keep laughing and loving, it's from the source of creation. Take nothing seriously. All things are light and easy because I am eternal love, nonthought spirit essence. Something incomprehensible because it is permanent and everlasting!

The more we proclaim things to be a fact, and categorize things in static modes of perception, the more we are convinced this reality is real, and the more we take it all very seriously!

This is why tribal people have such a spiritual nature, while the typical scientist is almost void of any spirituality whatsoever, two types of human lifestyles that express themselves in opposite ways. Tribal people see this world in a very spiritual sense and typically believe that everything in this world and the universe are connected. While scientist usually believes in cosmic accidents, and randomness in things, and try to devise mathematic formulas that fit their very closed perception of the universe. The tribal people gain insight by going into spiritual realms by having out-of-body experiences, drug-induced alternate states of mind, and through the use of vibrations through the use of drums, dance, song, etc. I'm not saying that all tribal spiritual insight is accurate, but I do believe that they are far more spiritually aware than most scientists, and even your typical religious Westerner. I think that some forms of spiritual perception can, at times in some cases, be a form of subtle science. Instead of ignoring these realms, I believe that science should explore them further. Some of the most delicate realities of this Universe may exist there.

I don't mean to keep bashing religion; I'm just stating the facts so that anyone who wishes can make a decision on their beliefs based on knowledge and research. I honestly do love all people. I just wish they would come from under the veil of darkness that covers their eyes and embrace all humanity instead, and stop allowing their beliefs to divide us.

Money, another major divider is nothing but numbers; it has no value besides what our beliefs project onto it. It seems to me that the Universe is a vibration-sound computer or an algorithm-quantifying machine that works with vibration and sound. This does not mean it is anything like any known computer in existence. But instead may suggest that there is an intelligent source to all that is. At times we have a terrible dream, and at other times it is filled with the joy and love we crave; remember these moments too. I love the holidays when everyone gets together and shares happy times. But we don't have to wait for the holidays. It can be done all the time. It's all a state of mind.

Human beings are like a single entity split into 8 billion smaller parts to express all the varying degrees and qualities of what it is to be human. To eliminate any part of us or any group is to lose part of our human family. It is holographic; we are all reflections of each other, reflecting varying characteristics of the whole.

Life is an exchange of energy through symbols, or coding expressed through matter! Since I observe this as the truth, then I would like to add an idea, which I call the popcorn to effect. We all have watched popcorn pop, and with our ears, we listen to its slow start. One kernel pops, then two seconds later, the popping is more and more rapid, and before you know it, the popping becomes constant until it's done. This is how I see the progression of the awakening of humanity to self-awareness and knowledge of oneself. With each person that becomes aware, it becomes easier and more accessible for that next person, until all those who were ever going to pop have done so. All those who have not popped are like the kernels left over at the bottom of the popcorn bag. It's either that they needed more time, or they were never going to pop at all.

CHAPTER 8

ORIGINAL SIN ORIGINAL EVIL

According to biblical scripture, there was a split between the union of God and Lucifer. They went from being best buddies to the most notorious enemies in known literature. First of all, I see this view of God as extraordinarily limited, and I do not know this story as being based on truth. Since so much of our religion and morale is based on this story, I will engage this topic in my critique. I see Lucifer as a symbol of lack within the biblical almighty God himself. I believe this because, according to the bible, God demanded that Lucifer worship man, which Lucifer refused to do. It seems as if God was into the caste system, a system of hierarchy! This is very odd to me; I couldn't imagine such a being who could create and do anything in all of the existence demand that one entity worships another. Especially an entity other than himself or herself!

Since Lucifer did not worship man, God engaged in a war with Lucifer and cast him out of heaven. I'm sorry to all who believe this story, but this sounds like nothing but a bunch of men talking, trying to get the minds of humanity to believe in their version of God, fashioned after their image. What Lucifer wanted was to stay close to God, having been created first, and not forced to worship God's later creation called man. Lucifer God's most beautiful angel at this time thought that man should worship him instead, but he did not demand it.

So what happened was that Lucifer expressed many human emotions, such as jealousy, and rage. Imagine that your best friend, and possible parent, created something inferior to yourself, then made you worship it. Would you not rebel? Would you not be angry? Would you not express the human emotions of jealousy, anger, rage, resentment, and very likely a broken heart? Lucifer expressed these very human emotions at this time, and because of it was denied the presence of God, his love, his best friend, and more, and was sent to the pits of hell. How unfortunate! I see this as the original Biblical Lack, which is a God that is limited to a mere human range of emotional and creative ability. Hence are born the evils as we know them today. All of the darkness in existence according to religion, are to be blamed on an entity that we can't even see, that we now call Satan, formerly Lucifer, God's most beautiful and adored angel.

Because of this little story, which is undoubtedly based on an older, possibly Sumerian tale, we are led to believe that Lucifer has been seeking revenge on humanity since the time of this occurrence. The reason why I think this story is damaging is that it leads people to blame their behavior on an entity called Lucifer, aka "Satan," instead of taking responsibility for you're their actions, and thus changing the self.

I believe Lack to be the first evil ever because it left an emptiness to be filled. This Biblical God could not fill Lucifer's void, according to the result of the story. Hypothetically if this story was true, these human feelings of lack repeat themselves over and over again in every way humanly possible. So, in essence, we live in Lucifer's pain because of what he experienced, and the revenge he is exacting on humankind. Hence Lucifer is the architect of our universe. Following this event, our Universe is based on lack, his symbols, his architecture his chronology. I believe this hypothetical situation mirrors reality because it is the paradigm we live in. Good thing this situation is temporary because we come from a whole source and will return to that. Once we become self-aware, balance our minds, and fill in our voids. Love is an excellent filler for that void. But it's a bit more complicated than just sending and receiving love because people see love from innumerable perspectives.

I believe original sin can also be in ways connected to the worship of the religious view of God. What I see as the original sin is, in my opinion, man's disconnect from nature. The symbolic clothing of oneself! No other living creature on earth clothes themselves besides Humans. When you put clothes on, you are symbolically saying you are separate from nature and all other human beings. Think about it, when you clothe yourself, you are effectively blocking yourself and all your intimate parts from being seen by others. Why is this done? As far as I know, it is currently done because we are trained this way from birth. We are taught to hide our genitalia from others, while other creatures walk about with no awareness that they should be hiding their genitalia at all. This is all kind of funny to me. So why do humans do this?

Maybe we can find the answers by looking at ourselves and discovering what effect clothing has on ourselves and others around us. So when we clothe ourselves, it leaves people around us with an unknown of what we look like under our clothes. In a sense, we are hiding from each other. When you are hiding, it means you are insecure about something. So it looks as if we are trained to have insecurities about ourselves. So clothing ourselves is all about security, which ironically causes insecurities in the first place. Clothing is the prison that guards our insecurities.

Clothing is, for the most part, worn to guard our insecurities dealing with our bodies, genitalia, and, most importantly, our psychosexuality. I believe this is the valid reason why Adam and Eve hid and clothed themselves. It was a psychosexual insecurity. I'm not sure what may have caused the psychosexual uncertainty dealing with the event that occurred involving the apple, and the tree of knowledge of good and evil, because these are just metaphors. Maybe the exact cause will be revealed to me one day.

This psychosexual insecurity has led us to believe that we have dirty parts that we should be ashamed of. Well I say what we call our private parts were never dirty, and never will be, and we should not be ashamed of any part of our body whatsoever. I believe that becoming ashamed of what we call our private parts and clothing ourselves, for this reason, is the real original sin. It opens the door to deception and dishonesty because our psyche is in the mode of constant hiding. We have been conditioned to be comfortable in hiding because we are penalized for disrobing. It's more honest to be naked!

Without this shame of our private parts or genitalia, and every thought concerning them, this style of civilization that we see today would not exist. We have alienated parts of ourselves by clothing ourselves. It was never meant to be this way. This is one of the major factors that hold this grand delusion we call our society together. The barrier between you and your sex or nakedness polarizes the individual with the use of psychosexual insecurities and the compartmentalization of the mind. When the mind is compartmentalized, which is in all conditioned people, the brain has fear blocks that keep parts of you separate so that you cannot access your true self, and many other parts of yourself for that matter, including love and self-love.

Imagine a world where archetypical insecurities didn't occur. This would mean that people didn't wear clothing because they would have nothing to be ashamed of. If people have no fear or shame, then they would also not need validation. Our society programs our minds from birth to seek and receive validation to feel like a being worthy of living in this world. If you are not validated, then it seems natural that you feel worthless, much like the poor and homeless we see around the world. They are indeed not validated.

The need for validation also keeps us acting like clones of each other because we lose the courage to express individuality for fear of not being validated, and becomes very easy to control. The controllers of the human species, whoever they are, have been very successful at squaring the circle. The Human consciousness is metaphorically a circle that can expand infinitely in all directions, but the "masters" of the Human race have been extraordinarily successful at putting the mind of nearly an entire species into a box.

I believe this started with what I call the original sin, which is man dawning on clothes because of his perceived insecurities. This was taken further with religion, a plethora of established institutions, and frames of thought. Notice that the phrase "frame of thought" there is the word "frame" which is synonymous with the word boundary or box. So, in essence, boxes of thought. For people to step out of their boxes of thinking, they must first become aware of them. If you never know that you are in a box, why would you ever try to get out? Naturally, you won't!

The way the world is contrived is so simple because it leads people into control through systems of belief, the beliefs themselves can be complicated, but yet they are still beliefs. These beliefs separate us from nature, ourselves, and each other. We are meant to harmonize with nature, much like all other living things. Native cultures do a decent job at this. I am not anti-technology or anything like that, but technology doesn't have to be anti-nature. Besides, our technology would be far more efficient if it was following nature. Rather than something that tries to conquer nature!

I think the progression of our human species is far less about evolving into something else entirely different from the way evolutionists see things, but instead now more privy to humanity simply waking up to who we already are. Part of this living universe!

CHAPTER 9

THE BLIND MAN

Man or Human beings, in general, are blind in many ways, metaphorically speaking. I find it very disturbing that many people cannot recognize other people as mere people just as themselves. Some seek to find any way they can to devalue the life of another. As far as I know, there are no set criteria for human superiority or inferiority that can be proved, especially on broad racial levels.

From my view, it seems so simple and apparent that we as humans would recognize that any other person, no matter what they look like, or what they believe, would have all the things in common that make us human. Such as the need for love, food, and shelter, and to simply be happy and get along in this world! But in this world, our human species regularly deny each other these bare essentials. It seems as if we are a broken species. But anything broken can be fixed, and they are a lot of people who have taken their blinders off and are working through love to fix this world.

Racism is not about the skin or even facial features. It's about the fact that there is a difference no matter how small it stands out compared to what another may view as the status quo of standard appearance, which makes what is seen by a majority within any given population the "standard type." Man can choose to split our species up into a thousand more races based on body types, facial features, and skin tone, or we could simply say they are no races at all, and a human is a human, and that's all that matters. In the end, all the racial talk doesn't mean much because it has to be based on the decisions of a man, and how he is willing to divide up his species into ever smaller parts.

These differences could have been called anything or nothing at all; some men chose to call these small differences race. It's all a part of the box that we are conditioned to live in. Though the word race has no place in the human family, there is no biological basis for this word in our species; it is strictly a social-cultural phenomenon. We project our beliefs onto others, and things are not necessarily the way they seem.

Categories or means of separating people within the human species have been manufactured and perfected with the use of fear for the past few hundred years. The concern is conditioned into each new generation, from parent to child. Children are not born with their views, and they have to be taught. Parents teach their children through their behavior, and word of mouth to dehumanize others, and to deny the existence of our shared human qualities. We are told to pay close attention to the stereotypes of a particular group and to look over their overwhelming similarities. So based on stereotypical differences, leads some people to believe that another human being feels different internally and somehow slightly alien to the way they think. When in fact, this is entirely false, I am quite sure that no matter what your body looks like, the shape and color of your features are not going to make you feel any different in your body. For example, if your body was to change suddenly, your skin became lighter or darker, and the shape of the features on your face were to change, I am 100% sure you would still feel like you. This happens naturally to some people when all of a sudden, a case of vitiligo has changed their skin color, and there are thousands of other reasons why a person's body can change appearance. A straightforward one being a suntan!

Those of us with racist "beliefs," (I would like to emphasize the word "beliefs" because racist beliefs are never based on facts), see those who may not be as close in likeness based on appearance as him or herself, to be subhuman, or not quite as human, and therefore subject to oppression and discrimination.

For instance, can you or anyone else feel their skin color? Or, more accurately, your "hue" because human skin doesn't come in different colors, it's just the concentration of melanin you have in your skin, giving you a different hue than the next person. It's something like pressing a pencil to a piece of paper and changing the pressure you apply to it. You will see that the hue changes, but it is all the same color. Can anyone feel the color of their eyes or the shape of their facial features? I don't think so. No matter what your features look like, or are shaped like, you would still feel the same in your body. The flesh does not give off different sensory information according to its shape and color.

So, in conclusion, you will feel the same no matter what your body looked like. So there is no right or logical reason to treat each other differently based on such extremely superficial terms. All human flesh is fundamentally the same. It all acts and feels like human flesh. You have to be taught how to feel black or white or any other race for that matter. You are not born knowing these modes of behavior. They are strictly cultural and not biological. We are all the same animal.

We treat each other as if we are limited to what our flesh looks like. Factually the flesh has nothing to do with what the human being feels in their heart or mind. When people say that they are a black person or a white person, they are, in essence, meaning that they are their flesh and that they are limited to that behavior paradigm of the way a black or white person is supposed to behave. It's a preoccupation with what the body looks like. But this is backward, heart and mind come first because this is where you are resonating from, not your flesh, which is simply your exterior appearance. We are spiritual, feeling beings with fleshly bodies, not the other way around.

Contrary to the belief that the government and media outlets are doing their part to do away with racism, they enforce it severely. Our government and the media push it as a significant force to remind us to pay attention to our bodies continually. We are told to define ourselves by our small differences, then to make judgments based on this. From early childhood, we are forced to fill out forms that divide us by race, so that our information found on this could be divided up and checked by a census bureau. Then reported back to us via the evening news, thus making what is unreal seemingly real. When, in fact, race has no basis in biological science! It's not only the evening news that reinforces the belief in racial division, it's TV shows, sitcoms, comedy sketches, and movies. They all play their part in the ceaseless attempts at division, (dividing our family). They reinforce stereotypes to no end and cast only according to specific stereotypes. It's overwhelming and pervasive. Even a person who is not a racist has trouble seeing through this, and just by default, may have some of their beliefs molded by the things they watch.

This has caused a tragic recycling of poisonous beliefs that only reinforce fear, separation, and violence within the human species, and hinder the acceleration of progress when culture and new ideas mix. Our media and the governing body need to do a much better job of educating people on our differences and show simply that these differences are no need for alarm, violence, and separation. With all the emphasis that is placed on education, it seems as if the world of education ultimately overlooks the need to educate people on who we are, and our relationship with one another. In the classroom, engaging one another is discouraged in most cases, and we are told to pay attention to the single authority figure, the teacher. So we are taught in terms of hierarchy and division. We are made to fiercely compete with one another to see who comes out on top. We are taught words like survival of the fittest, instead of the fact that we should be far more compassionate and sharing towards one another. We leave our fellow man to turmoil in his shortcomings. This is our conditioning since birth.

With all this said, I think, for the most part, the human being is innocent and operates something like a computer (on the most basic level of low awareness), which most of us are in. You only get out of the individual what you put in, for the most part. We are born into this world's social environment, which is broken and unbalanced, and become cracked and unbalanced because the people already here are broken and unstable. Though we are all capable of critical thinking, many people, or most people, do not engage in it when it comes to analyzing their world environment and the belief systems that they are taught. More people need to question the things they think they know. Especially what they believe they are confident of because it could be bigotry.

In the grand scheme of things, humans live out their existence more or less mechanically. I firmly believe that the clashing of cultures over the past couple of hundred years may have a unifying effect or a merger of some sort. Instead of being solely due to the cause of individual agendas, I believed that there is a more astronomical unnamed phenomenon at work that is uniting our species after untold millennia of isolation between tribes and nations. As this happens, we are sometimes taken aback by the differences we see between us. For fear of the unknown, we sometimes meet these differences with ignorance and violence. But the more intelligent of us or simply, the less severely brainwashed, meet these differences with understanding and an open mind and try to take note of what a different culture may have to offer in terms of ideas, skills, cosmology, and spirituality. So, in turn, ideas and love are exchanged.

I think it could be safely said that most people are curious about people who have different cultures and may look different from themselves. And it seems to turn out that the more you think people are different, the more you find out that they are the same.

The conquering of other nations, colonialism, immigration, and the slave trade, was never solely for the agendas of the individuals because those agendas are short-term. The long-term repercussions of those actions outlive the immediate agendas for thousands of years, or as long as they are people alive. Every action has an equal and opposite reaction. The moving around of millions of people, and the mixing of people who were once foreign to each other affects individuals in ways that are not predicted by the perpetrators of short-term agendas, or even long-term agendas for that matter. I believe there is a more astronomical phenomenon at hand, which is the uniting of our human species.

Think of how you met that new kid in your class for the first time. At first, you may be shy and hesitant, and maybe even act out violently. But as soon as you start to realize that the individual is just like you, and has so much in common, all the shyness and animosity quickly fade away. This is happening on a far more massive scale, and there is some resistance. Still, sooner or later, people will get over the resistance and shyness syndrome we experience when a new kid comes around, and will quickly embrace each other, forgetting the old feelings of uncertainty and shyness we see in the new kid syndrome cases. Once you feel that you know them, then that's it, those feelings usually do not reverse themselves.

The attraction human beings have toward one another has manifested over time as the coming together of the world's populations. There is evidence to suggest that in our ancient past, we had a civilization that shared a common language worldwide. Through the course of human history, we may be going through a great deal of build-up to high technology and worldwide governments, only to be destroyed by natural or even human-made phenomena. In turn, we become stone-aged people again. Destruction, dispersion, isolation, and reuniting may be an ongoing cycle for humanity, especially if we do not learn from our past and stand together. This is my hypothesis, and I believe a lot of evidence backs this up. It just has to be brought to the forefront of attention.

With this being said I believe that the attraction that all humans have for one another, maybe somewhere in our DNA or the subconscious or unconscious, is definitely beyond the individual's control no matter what their beliefs are. It's never about the moment, but what the moments add up to later and become the big picture. It was never solely about the conquerors, colonialists, slave traders, and immigrants, but the product of the collective society; therefore, humanity as a whole will run its natural course because we are dealing with spiritual beings, and not just bodies waiting to be manipulated.

All supremacist groups are driven by fear and anger, not love. If supremacists went about spreading their superiority based on the fact that they can achieve supreme love for their fellow human brothers and sisters, then maybe they would be believed. But instead, they are shunned by most because they do the exact opposite. They alienate people with their messages of hate and division. Living in constant hate is a degenerative state that can cause mental illness. Hate groups are nothing to be feared because they are already frightened; this is why they openly express such aggressive behavior to gain power over the psyche of another. The gaining of power through aggressive manners shows that the initiator of aggressive energy is already weak, and needs you to fear them to gain the control they need. Without your fear, they are nothing, powerless.

This book has been revised by myself, and I would like to add to these predilections as I thought about the mind of an overlord, and their largely reptilian way of thinking and ruling. I have come to realize that it is reptilian primarily to be racist. Human beings are mammalian and geared more toward empathy. But we are still reptilian beneath our more complex exterior mammalian brain. I have heard the reptilian brain in humans is called the "R Complex" or the Basal Ganglia. Hierarchy, ritual, cold empathy, or non-empathic behavior are reptilian attributes within humankind. Our elite overloads knew how to conquer and divide us since the beginning by giving us their reptilian minds and their reptilian way of thinking. As warm-blooded empathic human beings, we can break out of this hypnosis after realizing that this has been the collective spell for eons.

CHAPTER 10

LACK AND THE HUMAN'S ANIMALISTIC INSTINCT TO HOARD

Billionaires seem to have complete and total abundance with nothing to worry about. But instead of allowing their billions to trickle down to benefit the 98% - 99% of the remaining human population, it seems as if the hoarders try to gain as much material wealth as possible. Hoarding can be done by individuals, corporations, governments, and even religious institutions. The Vatican itself has acquired more wealth than any other single institution or government for that matter. More than enough to feed the world many times over, and yet the wealth is just sat on. The only ethical way to be a billionaire is to allow your excess income to fund the assets of those in need. A proper visualization of this concept is to picture Niagara Falls with all its abundance of water cascading into the ravine below, which would be dry without it. The current attitude of most billionaires is like a flowing river that has been dammed, leaving a once-flowing river valley dry and thirsty.

All people seem to hoard to some extent, but we see this phenomenon expressed most strongly by the super-elite. Who hold at least 60% of the world's wealth, while being only 1-2 % of the world's population. Why do we see this type of extreme hoarding while billions of people still live at sub-poverty levels?

Has this human instinct gone extremely haywire? Do some humans have such a strong survival instinct as to feel indifference in the presence of an impoverished human being? Perhaps this hoarding is driven by a primordial fear of lack. The individual may be in a state of mind that you do see and care about others suffering around them but looks the other way because you rather that it be them and not you, in turn, continuing to hoard excessively. Or, for some people, it may be a legitimate evil, and they may wish for the death and suffering of others.

In my opinion, those who are evil have the most fear, because they put the most effort into condoning the death and suffering of others so that they have less competition for wealth and resources. This is the world we live in! Most people have been trained to go on a rat race for a lifelong quest for wealth and resources, so it's challenging to see the grand scheme from the inside. I do not believe there is ever a good reason for anyone to hoard billions of dollars while they are billions of people suffering due to substandard living conditions. Though wealth is a beautiful thing to have, I do not frown upon wealth and abundance itself. I only frown upon the indifference of the super-wealthy project when it comes to the suffering of others.

Hoarding may also have some psychological, and energetic effects on the individual and his or her surroundings. A hoarder is typically a collector of a massive amount of human-made objects that are perceived to be worth something. So what happens when someone surrounds themselves with an enormous amount of human-made items? Typically things such as nature and spirituality become less evident to them, and these things quickly lose their importance. The more material wealth a person gains, the more they usually seem to separate themselves from nature and the need for a Godly presence or Creator force. The matter acts like a barrier or physical shield because wealth usually surrounds the everyday scenarios of the individual's life. The thoughts of a billionaire seem to be more focused on the amount of material wealth that can be gained, rather than nature, spirituality, and the well-being of other individuals. Though they are wealthy people who are an exception to these paradigms, live noble, beautiful lives, and are aware of the world around them.

SEXUALITY

For many people of religious and non-religious backgrounds alike, sexuality can be a very sensitive topic. We as human beings know very little about what drives our sexuality, and at times, any single individual's sexuality can be severely complex. So I will just throw a straightforward theory out there, which shouldn't offend any mature person. I will look at sexuality from a male and female energy exchange point of view. When a male desires a female, he is seeking feminine energy to balance out his masculine energy so that he may harmonize himself. When it comes to a male who is seeking another male as a mate, it could be that he is seeking male energy because he may have a lot of female energy to balance out. The same goes for a female who is seeking a female; she may have a lot of masculine energy and may be seeking to balance it out with feminine energy.

The very same individuals who may have been seeking one kind of energy at first from a male or female may again seek different energy of another sex once they have gotten the energy they were seeking from the initial sex. This may be why sexuality is so complicated because energy exchange may be complicated, and we may continually be looking for balance on certain levels, which we are not consciously aware of. Another factor for some individuals may be that their physical sex may not match their energetic and or mental sex. So someone may physically be a male, but grow up believing that he is born into the wrong sex and prefer to be a female, and the same thing happens to some people who are physically female. This may all boil down to female and male energy balance.

I'm not trying to agree or disagree with anyone or any particular set of beliefs. This is just how my mind best perceives the complex nature of human sexuality. Scientist is also discovering that some 50% or more of species on this planet may have a sexuality that does not fit the standards of what we see as typical sexual behavior between male and female. Most plant species are both male and female. Many worms and fish can change sex or are hermaphrodites, and female hyenas have a pseudo penis and will mount their opponents to show sexual dominance, etc., the list goes on and on.

THE CAUSES OF ABERRANT SEXUAL BEHAVIOR

For one thing, I can not say that I am 100% sure about the causes of all sexually deviant behavior such as rape and molestation. But I will give my theories as to what I think some of the reasons may be, though they can vary in any given individual because all are unique in circumstance and upbringing. My take on what I observe with rapists is that they may be starving for sexual energy; this may be why they act out so aggressively. For instance, when you are starving, you may hunt your food and eat it very fast and aggressively.

The same may go for someone who is starving sexually, since all people may require sexual energy in the same way they need food, the sexually starved person may rape aggressively because they were starving for that energy. I know these anecdotes are very general, but they are also pragmatic.

A child molester may be experiencing a sort of sexual retardation. It seems as if their sexual age is stuck at the age or close to that of their victims. So the child molester will always seek energy at the level the individual's age is frozen. The trauma that caused their sexual retardation may lie somewhere in their past. If those moments can be pulled to the surface and dealt with, the condition they are suffering from can be cured so that they can live balanced and healthy lives without harming others. Without this being done, these people go on to harm hundreds of child victims in a single lifetime.

When it comes to other aberrant behavior, we lack knowledge as to why people commit particular harmful actions. I believe that all you have to do is follow the pattern of thoughts as the result of a specific phenomenon in any person's life leading up to the occurrence of any behavior good or bad, but this is especially helpful in aberrant behavior. In doing this, you can cure the person and treat them for their behavior, rather than dole out a system of punishment that has nothing to do with eliminating the behavior in the individual. For instance, if a person is a thief, the person is thrown in jail instead of being questioned in a way that we can prevent this kind of behavior from reoccurring, and in turn, manipulate society in a way so that people won't feel the need to steal. Logs of accounts given by criminals on their behavior should be journaled. We can then understand the flaws within society and the individuals so that we may be able to eliminate criminal behavior, rather than figuring out the best way to punish these people, which is archaic and feudalistic, and thoughtless in practice. But if we discovered that behaviors can not be cured and may instead have a maladaptive genetic factor, then particular dangerous individuals such as rapists, murderers, and child molesters must be removed from society permanently!

Our society appears modern because of technology. But we are no more advanced in terms of our moral behavior and the way our society is run, than our feudalistic predecessors in the way we think. Western civilization as a whole is still steeped in profound ignorance, and people lack the will to think for themselves and instead love to follow the leader. Until we look at ourselves in the mirror and think critically about ourselves and our surroundings, these problems will continue to persist. These are problems that are simply symptoms of a broken society filled with broken individuals.

CHAPTER 11

SELF ACTUATED SOLUTIONS

From what I observed by merely living and learning, the solutions to the mess we are in, are where ever we look. But we look and do not see because we have developed knowledge filters in the mind that impair us from embracing secure solutions. So instead, we seek the status quo and warn each other never to stray from it. The status quo today is what we see as normal in society.

We wake up to an alarm clock, quickly obeying the demands of time, with no questions about its position of authority, it tugs and pulls at our being as it leads us in an everlasting race until we are old in our graves and have forgotten where all the time had gone.

Life is not meant to be a never-ending rat race to see who gets the most cheese. Because after it's all over, you will look back and see all the life that you forgot to live while you were racing along. Anytime you look into space or any point at a vast distance, you are essentially looking back in time, only because it takes light time to travel to your retina. So when we look out into space, we can never see the present. The present only exists in our immediate proximity. So invariably, we are natural time travelers. Our bodies are the vehicles in which we time travel. We watch ourselves grow and age from infancy to death. Even our sunlight comes from the past, if our sun were to stop shining, we wouldn't even know it until 8 minutes later. I find that phenomenon amazing when it comes to putting the universe around us into perspective.

Lack ultimately is a state of mind. When we are searching for that thing that we are lacking, it turns out that what we are looking for is that state of mind that object or subject will give us, much like a drug. But since it is apparent that we go into these states of mind once we receive the things that we think we are lacking. Then it is certainly possible to manipulate our minds into these states without a tactile stimulus, but instead with the will of the individual alone.

For instance, if you have a negative thought, try thinking with the opposite view. Subject to this happening, your mind would start steamrolling in a different direction. Try it and see, and use your will to maintain a state of mind change. Though sometimes negative thoughts need our attention and should be analyzed, most of the time, they are irrational. As you become more aware, you will be able to use your discernment to determine this for yourself.

All emotion is energy. When we communicate with one another, we transfer this energy from one person to the other. For instance, when you are angry at someone, you are transferring your pinned-up energy to the individual who is the subject of anger and also to anyone else in the area, which is within sensory perception. This energy can manifest in the individuals it is transferred to, in many different ways. The individual may internalize this energy or immediately transfer it through aggressive or defensive behavior. All emotions given off by a person will have some sort of effect on those around them. Children are the most excellent examples of this being that they are so impressionable, and this emotional energy can manifest in an unlimited amount of unpredictable ways. Children are the most sensitive emotional energy barometers. We become much more callous as we grow older. Inevitably forgetting what it was like to be a child. To remember what it was like to be a child, you have to remember those emotions that you had when you were a child.

When it comes to human relationships, they seem to typically work the best when an individual feels they can let their guard down. When an individual lets their guard down, energy does not stay pinned up inside them, be it negative or positive energy, and you become a conduit where emotional energy can flow freely, which enhances communication and leads to greater understanding between individuals.

The most severe problem that we have as a human species is the fact that we feel disconnected from one another. Many people dissociate from others and the pain they feel due to any number of experiences. I don't think that anyone born goes through life without feeling a bit of pain, but sometimes that pain can be so unbearable to the individual, that it can cause the person to dissociate as a coping mechanism.

The more an individual can dissociate from their experiences. The more violent and callous individuals can potentially be when dealing with other people. For example, most of us dissociate entirely when we walk past someone homeless and hungry. In our society, it is entirely acceptable to do so; we have created a barrier between us and the emotional and physical plight the homeless individual may be going through. Therefore we do not empathize with the individual, and we keep on walking because we believe that we are intrinsically separate from that individual, but this is not true. Upon observation, we are already exchanging energy through any one of our senses. So we choose to either feel for that person or dissociate whether it's consciously or subconsciously.

Once you become aware of yourself and listen to what goes on inside your heart and mind, you become aware of your decisions that affect your emotional state of being. You will begin to know yourself for the first time on a level that you could not have previously imagined.

OUR COLLECTIVE UNIVERSAL EXPERIENCE

We all must believe that our world is a good and beautiful place to be in because our reality is only as real as our perception of it. If you are standing idly by waiting for things to become perfect, you may wait ever. You must attempt to perfect your mind before you ever see the world as being perfect or healed. When you heal your mind, you heal the world. But you do not want to be an accomplished perfectionist, because an accomplished does not look for imperfections to perfect, but instead looks over flaws and only acknowledges what is already perfect. So improving the mind must be an everlasting journey because there are no limits to its potential, and it can always get better.

The Universe itself is a manifestation of our shared perception. At some level, we have access to this shared collective of mind. Some quantum physicists call it the unified field. Separation is only a condition or experience, but is categorically false on an energetic quantum level, for there is no true separation from the source. We are one mind split into infinite time, space, organic, and nonorganic material scenarios.

Life is the exchange of energy through symbols or coding expressed through matter, light, and vibration, such as sound decoded by the body and mind.

The awakening of human awareness is always more accessible for the next person after someone has previously accomplished it. Once you hear one popcorn pop, you know for sure you will hear a cascade of future popping kernels. The popcorn effect generally applies to any human accomplishment, such as track and field, technology, sports, etc. Everything is a thought form.

THE GRAND RECYCLING

In this universe, we live in; something always seems to consume and take over something else. It's a never-ending cycle! Is there an ultimate goal for all of this recycling of matter? I am not sure, but the cannibalism of matter seems to be an ongoing phenomenon, at least from the human perspective.

Matter is energy, and since there is truly no empty space, then all energy is one. The more massive and dense the object or matter is, the more energy it holds and can consume. The more energy something must consume, the more significant its apparent lack.

For example, the larger the animal, the more it eats—the more massive or more dense the celestial object, the more significant its gravity and energetic need. So gravity may just be a tool used by celestial bodies to pull energy into their centers.

All living things must consume the right amount of elements to feed their biological life force. Like a puzzle whose every piece has its place, the body integrates the elements of the universe into the body in perfect amounts. (when a person eats correctly). This is a constant throughout the life form's life span.

But Biological life forms cannot contain or hold onto energy forever. The energy is continuously dissipating into the universe. So it's an endless search for more energy.

I presume this to also be true with celestial bodies, especially self-heated balls of energy we call stars and Jovian planets. Even smaller celestial objects, for the most part, are part of a more extensive energy system such as solar systems and galaxies. It all works in scale. The Universe is truly holographic.

Lightning dissipates into sound and heat, two different forms of the same energy. Energy is never destroyed; it only builds up again at another point in space and time until its next release. The higher the temperature of the subject, the more energy it must use to maintain that temperature. Hence warm-blooded creatures need to eat more than cold-blooded animals to maintain homeostasis.

We also discussed earlier that everything is vibrating, the science of cymatics. In conclusion, the theory that I was able to come up with dealing with all matter and vibration, and the integration of different energies and frequencies, is that the hotter the object, the faster it is vibrating, this is why something hot may destroy something cooler. The more heated object has a higher frequency of vibration than the colder object, so in turn, it is destroyed by the faster-vibrating object. Stellar objects or suns may be entities vibrating at a very high speed that we feel as heat and light energy. They may bend time and space around themselves; this is why we revolve around these objects, with time on Earth, depending on the speed at which it revolves around its sun. But this subject alone could be the beginning of a whole new book.

CONCLUSION

Dear reader, I am quite aware that I may have put a lot of information backed up by my thoughts and research onto the world stage. Some may describe it as being controversial. Others may see it as a beneficial life-changing, goal-orienting, mind-awakening, inspiring conglomerate of information. To those with minds who are open to new information, I congratulate you. The open mind is adaptable, analyzing the world around them to come up with the most efficient scenarios to benefit themselves and others around them. The closed mind is the exact opposite. The closed mind is like a stagnant cesspool of overgrown algae choking away all sources of fresh flow. It is rigid and self-serving with very little room for change or adaptation. Hence survival for this person when a new paradigm arises is tough.

To those who are driven by fear, and want to cause harm to others to preserve their fear-driven mentality and beliefs, I hope these words written here bring you peace and comfort. If you wish to harm another, look in the mirror and ask yourself why, and you will see that the anger that you are expressing outward is just a fear mentality that feels threatened and wants to keep itself in place by eliminating its threats. But if that state of mind were worth keeping, it would not be so easily shaken by the mundane. Who you genuinely fear is yourself, it is only your mind that tortures you. All thoughts are contained within the confines of your own skull. So bring the love of yourself and others into your mind, and soothe away whatever negativity that binds you.

WORKS CITED

Bible: New international version. Biblica, 1973, 1978, 1984

Lost Civilizations Sumer Cities of Eden. Time Life Medical 1999

Willis Barnstone, The Other Bible. Harper Collins 1984

Printed in the United States
by Baker & Taylor Publisher Services